Table of Contents

Introduction

How to Install and Repair Gas Lines

A DIY Guide to Home Gas Appliance Installation, Maintenance, and Troubleshooting for Homeowners and Contractors

Cole Bennett

Chapter 12: Future of Home Gas Systems
- Smart Gas Meters
- Renewable Natural Gas
- Hybrid Systems

Conclusion

Introduction

Picture this: you're preparing dinner for your family, and as you turn on your gas stove, you notice a faint smell of gas. Your heart races as you realize there might be a leak somewhere in your gas line. Panic sets in as you wonder about the safety of your loved ones and your home. Sound familiar?

As a homeowner, your gas line system is one of the most critical yet often overlooked components of your house. It provides heat, powers appliances, and enhances your quality of life. However, like any utility system, gas lines can develop issues over time, and when ignored, they can cause significant inconvenience and pose serious safety risks.

That's where "How to Install and Repair Gas Lines - Stoves, Fireplaces, Grills, and Safety" comes in. This comprehensive guide is designed to empower you with the knowledge and skills necessary to tackle common gas line problems confidently. Whether you're dealing with a new installation, a suspected leak, a malfunctioning appliance, or routine maintenance, this book will walk you through the step-by-step process of handling these tasks safely and effectively.

But why should you bother learning how to install and repair gas lines yourself? For starters, professional gas line services can be costly. By mastering the skills outlined in this guide, you can save significant amounts of money on installation and repair costs over time. Plus, you'll have the peace of mind that comes from knowing you can handle gas line issues promptly and safely, without having to wait for a technician during an emergency.

In this book, we'll cover everything from the basics of gas line systems and common problems to essential safety precautions and tools you'll need to work with gas lines properly. We'll demystify the workings of your home's gas system, helping you understand the function and importance of each component, so you can diagnose issues and perform installations with confidence.

But don't worry – we won't overwhelm you with technical jargon or complex theories. Our goal is to make gas line installation and repair accessible to everyone, regardless of their prior experience. With clear, concise instructions, helpful diagrams, and practical tips, you'll feel like you have an expert guiding you through every procedure.

So, whether you're a DIY enthusiast looking to expand your home maintenance skills, or simply a homeowner who wants to be prepared for any gas-related issue, "How to Install and Repair Gas Lines" is your ultimate resource. By investing in this guide, you're not just buying a book – you're investing in the safety and efficiency of your home's gas system for years to come.

Are you ready to take control of your gas lines and become your own household gas expert? Let's embark on this empowering journey together and ensure the safety and functionality of your home's gas system!

Chapter 1
Introduction to Gas Line Systems

Types of Gas Used in Homes

Picture this: you're preparing to upgrade your old electric stove to a sleek, new gas range. As you browse through options, you suddenly realize you're not sure what type of gas your home uses. Is it natural gas or propane? The uncertainty sets in, and you find yourself overwhelmed by the technical jargon and safety considerations. Sound familiar?

As a homeowner, understanding the types of gas used in your home is the foundation of safe and efficient gas line management. It's not just about choosing the right appliances; it's about ensuring the safety of your family and the optimal performance of your gas-powered systems.

That's where our deep dive into the types of gas used in homes comes in. In this section, we'll demystify the world of residential gas types, empowering you with the knowledge to make informed decisions about your home's gas systems.

Types of Gas Commonly Used in Homes

There are two primary types of gas used in residential settings: natural gas and propane (also known as Liquefied Petroleum Gas or LPG). Each has its own unique characteristics, advantages, and considerations. Let's explore them in detail:

1. Natural Gas

Natural gas is the most common type of gas used in homes across the United States. It's a naturally occurring fossil fuel that's primarily composed of methane.

How Natural Gas Works:
Natural gas is delivered to homes through underground pipelines maintained by local utility companies. When you turn on a gas appliance, the gas flows from the main line into your home's gas pipes and to the appliance, where it's ignited to produce heat.

Advantages of Natural Gas:
- Cost-effective: Generally cheaper than propane and electricity
- Convenient: Continuous supply through utility lines, no need for refills
- Clean-burning: Produces fewer emissions compared to other fossil fuels
- Versatile: Can be used for heating, cooking, and powering various appliances

Considerations:
- Availability: Not all areas have access to natural gas lines
- Installation costs: Initial setup of gas lines can be expensive
- Safety: Requires proper ventilation and regular system checks

2. Propane (LPG)

Propane is a byproduct of natural gas processing and crude oil refining. It's stored as a liquid under pressure but turns into a gas when released for use.

How Propane Works:
Propane is typically stored in tanks on the property. These tanks can be above or below ground. When you use a propane appliance, the liquid propane vaporizes and travels through pipes to the appliance, where it's burned to produce heat.

Advantages of Propane:

Portable: Can be used in areas without natural gas lines

- Energy-dense: Provides more energy per cubic foot than natural gas
- Environmentally friendly: Clean-burning with low carbon emissions
- Versatile: Used for heating, cooking, and powering various appliances

Considerations:

- Cost: Generally more expensive than natural gas
- Refills: Requires monitoring of tank levels and scheduling refills
- Storage: Needs space for tank installation on the property
- Safety: Proper tank maintenance and leak detection are crucial

Choosing the Right Gas for Your Home

The choice between natural gas and propane often depends on factors such as:

1. Availability: Check if natural gas lines are available in your area.
2. Cost: Compare long-term costs of both options in your region.
3. Appliance compatibility: Ensure your appliances are compatible with the chosen gas type.
4. Personal preference: Consider factors like environmental impact and energy independence.

Understanding the types of gas used in homes is crucial for anyone looking to work with gas lines. It affects everything from the tools you'll use to the safety precautions you'll need to take. By mastering this knowledge, you're taking the first step towards becoming proficient in gas line installation and repair.

In the following sections, we'll delve deeper into the components of gas line systems, safety considerations, and the practical skills you'll need to work confidently with both natural gas and propane systems. Are you ready to become your home's gas system expert? Let's continue this empowering journey together!

Basic Gas Line Components

Imagine walking into your basement and seeing a maze of pipes, valves, and fittings. You know they're part of your home's gas system, but what does each component do? How do they work together to safely deliver gas to your appliances? Understanding these basic gas line components is crucial for any homeowner or DIY enthusiast looking to work with gas systems.

When I first encountered my home's gas line system, I felt overwhelmed by its complexity. But as I learned about each component and its role, I gained confidence in maintaining and even repairing parts of the system. Let me share that knowledge with you, so you can approach your gas line system with the same confidence.

In this section, we'll break down the essential components of a gas line system, explain their functions, and provide troubleshooting tips for common issues. By the end, you'll have a comprehensive understanding of how your gas line system works from the meter to your appliances.

Key Components of a Gas Line System:

1. Gas Meter

The gas meter is where your home's gas system begins. It's typically located outside your house and is owned and maintained by your gas utility company.

Function:
- Measures the volume of gas used by your household
- Acts as the connection point between the main gas line and your home's system

Troubleshooting Tips:
- If you suspect a gas leak near the meter, leave the area immediately and call your gas company
- Never attempt to remove or adjust the meter yourself
- If you notice the meter running when all gas appliances are off, you may have a leak in your system

2. Main Shut-off Valve

Located near the gas meter, this valve is your first line of defense in case of a gas emergency.

Function:
- Allows you to quickly shut off the gas supply to your entire home

Troubleshooting Tips:
- Familiarize yourself with the location and operation of this valve before an emergency occurs
- If the valve is difficult to turn, do not force it - call a professional
- Consider installing an automatic shut-off valve for added safety

3. Pressure Regulator

This device is typically located near the meter and is crucial for safe gas distribution.

Function:
- Reduces the high pressure of gas from the main line to a safe, usable pressure for your home appliances

Troubleshooting Tips:
- If you smell gas or hear hissing near the regulator, evacuate and call your gas company immediately
- Fluctuating gas pressure in your appliances may indicate a faulty regulator
- Never attempt to adjust the regulator yourself - this requires professional expertise

4. Gas Piping

The network of pipes that distribute gas throughout your home.

Function:
- Safely transports gas from the meter to your appliances

Types:
- Black iron pipe: Most common, durable, and suitable for both indoor and outdoor use
- Flexible corrugated stainless steel tubing (CSST): Easier to install, but must be properly bonded and grounded
- Copper tubing: Used in some regions, but not universally approved for gas lines

Troubleshooting Tips:
- Regularly inspect visible pipes for signs of corrosion or damage
- If you smell gas near any pipes, evacuate and call for professional help immediately
- Never hang items from gas pipes or use them as support for other objects

5. Gas Valves

These are found at various points in your gas line system, including near appliances.

Function:
- Allow you to control the flow of gas to specific areas or appliances
- Provide a way to shut off gas to individual appliances for maintenance or in case of leaks

Types:
- Ball valves: Easy to use, quarter-turn operation
- Gate valves: Require multiple turns to fully open or close

Troubleshooting Tips:
- Ensure valves are fully open during normal operation - partially open valves can restrict gas flow
- If a valve is stuck or leaking, do not force it - call a professional
- Regularly exercise valves (turn them off and on) to prevent seizing

6. Fittings and Connectors

These components join sections of pipe and connect pipes to appliances.

Function:
- Create secure, leak-free connections between pipe sections and to appliances

Types:
- Threaded fittings: Used with rigid pipes
- Flare fittings: Often used with soft copper tubing
- Compression fittings: Used for some connections to appliances

Troubleshooting Tips:
- Regularly check fittings for signs of corrosion or looseness
- Use appropriate pipe dope or Teflon tape on threaded connections to ensure a proper seal
- If you smell gas near a fitting, do not attempt to tighten it yourself - call a professional

7. Sediment Traps

Also known as drip legs, these are typically installed near appliances.

Function:
- Collect any debris or condensation in the gas line, preventing it from entering and damaging appliances

Troubleshooting Tips:
- Ensure sediment traps are installed correctly - the tee fitting should be positioned with the branch facing downward
- Periodically check and clean sediment traps to maintain their effectiveness
- If you notice decreased gas flow to an appliance, a clogged sediment trap might be the culprit

8. Flexible Connectors

These are used to connect gas piping to movable appliances like stoves.

Function:
- Allow for slight movement of appliances without stressing rigid gas lines
- Provide an easy way to disconnect appliances for maintenance or replacement

Troubleshooting Tips:
- Replace flexible connectors every 10 years or if they show any signs of wear or damage
- Never reuse a flexible connector when installing a new appliance
- Ensure connectors are not twisted or kinked when installed

Understanding these basic gas line components is crucial for maintaining a safe and efficient gas system in your home. While many maintenance tasks should be left to professionals, this knowledge empowers you to identify potential issues early and make informed decisions about your gas system.

Remember, safety is paramount when dealing with gas lines. If you ever smell gas or suspect a leak, evacuate the area immediately and call your gas company or emergency services. Never attempt to repair a gas leak yourself.

In the next sections, we'll delve deeper into safe practices for working with gas lines, how to properly size and install new gas lines, and techniques for troubleshooting common gas line issues. Are you ready to become a gas line expert? Let's continue our journey into the world of home gas systems!

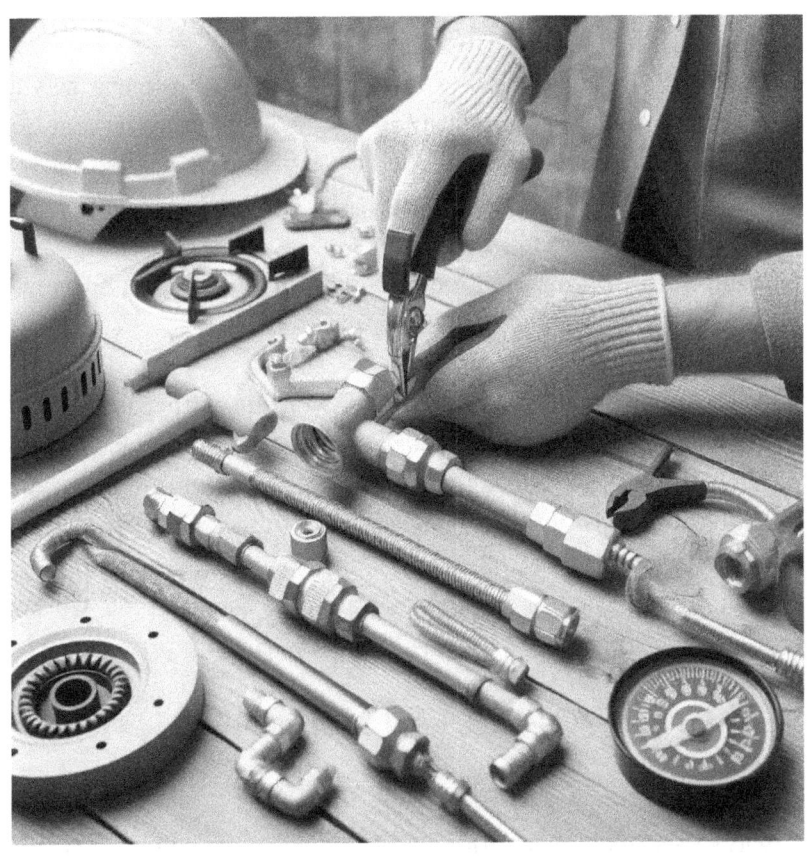

Safety Considerations and Regulations

Picture this: You're eager to install a new gas fireplace in your living room. You've watched a few online tutorials and feel confident in your DIY skills. But as you prepare to start, a nagging voice in your head asks, "Am I overlooking any crucial safety measures? What regulations do I need to follow?" This moment of hesitation could be the difference between a successful project and a potentially dangerous situation.

I've been there. When I first started working with gas lines, the sheer number of safety considerations and regulations seemed overwhelming. But I quickly learned that these rules aren't just red tape – they're vital safeguards that protect you, your family, and your property. Let me guide you through the essential safety considerations and regulations you need to know when working with gas lines.

Safety First: The Golden Rules of Gas Line Work

1. Never Assume It's Safe
Always treat gas lines with respect and caution. Even if you think a line is shut off or empty, proceed as if it's live and full.

Troubleshooting Tip: Before starting any work, double-check that the gas is off using a gas detector. Never rely solely on your sense of smell.

2. Proper Ventilation is Crucial
Gas can accumulate quickly in enclosed spaces, creating a dangerous situation.

Troubleshooting Tip: If you're working indoors, open windows and doors to create cross-ventilation. Use fans to circulate air if necessary.

3. Use the Right Tools
Using improper tools can lead to gas leaks or sparks that could ignite gas.

Troubleshooting Tip: Invest in high-quality, spark-free tools designed for gas line work. Never substitute with regular tools.

4. Always Check for Leaks
After any work on gas lines, thorough leak checking is non-negotiable.

Troubleshooting Tip: Use a mixture of dish soap and water to check for bubbles at connections, or invest in an electronic gas leak detector for more precise results.

5. Know When to Call a Professional
Some gas line tasks are simply too complex or risky for DIY. Know your limits.

Troubleshooting Tip: If you're ever in doubt about your ability to safely complete a gas line task, stop and call a licensed professional immediately.

Key Regulations and Standards

Understanding and following regulations is crucial for safe and legal gas line work. Here are some of the most important standards to be aware of:

1. National Fuel Gas Code (NFPA 54/ANSI Z223.1)
This comprehensive set of guidelines covers installation and operation of fuel gas piping systems, appliances, and related accessories.

Key Points:
- Specifies proper materials for gas piping
- Outlines correct installation methods
- Provides guidelines for testing and purging systems

Troubleshooting Tip: Always consult the latest version of this code before starting any gas line project. Local libraries often have copies available.

2. Local Building Codes
Many jurisdictions have additional requirements beyond national standards.

Key Points:
- May specify additional safety measures
- Often require permits for gas line work
- Can mandate professional installation for certain tasks

Troubleshooting Tip: Contact your local building department before starting any gas line project to ensure compliance with local codes.

3. Manufacturer's Instructions
Appliance and component manufacturers provide specific installation guidelines that must be followed.

Key Points:
- Detail proper connection methods
- Specify clearances and ventilation requirements
- Often required to maintain warranty coverage

Troubleshooting Tip: Keep all manufacturer's instructions in a safe place for future reference. They can be invaluable for troubleshooting issues later.

4. Gas Utility Company Regulations

Your local gas provider may have additional rules for working on gas lines.

Key Points:
- May restrict certain types of DIY work
- Often provide free safety inspections
- Can offer guidance on local regulations

Troubleshooting Tip: Build a relationship with your local gas utility. They can be an excellent resource for information and assistance.

Essential Safety Equipment

Having the right safety gear is crucial when working with gas lines. Here's what you should have on hand:

1. Gas Detector

A must-have for detecting gas leaks and ensuring a safe working environment.

Troubleshooting Tip: Test your gas detector regularly and replace batteries as needed. A non-functioning detector is worse than no detector at all.

2. Fire Extinguisher

Keep a properly rated fire extinguisher nearby whenever working on gas lines.

Troubleshooting Tip: Familiarize yourself with how to use the extinguisher before an emergency occurs. Check it regularly to ensure it's fully charged.

3. Personal Protective Equipment (PPE)
This includes safety glasses, work gloves, and protective clothing.

Troubleshooting Tip: Inspect your PPE before each use. Replace any damaged items immediately.

4. Proper Ventilation Equipment
Fans or blowers to ensure adequate air circulation in work areas.

Troubleshooting Tip: Position ventilation equipment to create a flow of fresh air across your work area and towards an exit.

5. First Aid Kit
Keep a well-stocked first aid kit easily accessible.

Troubleshooting Tip: Regularly check your first aid kit and replenish any used or expired items.

Emergency Procedures

Despite all precautions, emergencies can still occur. Knowing how to respond can save lives:

1. If You Smell Gas:
- Immediately evacuate the area
- Do not turn on or off any electrical devices, including lights
- If safe to do so, turn off the main gas valve
- Once outside, call your gas company or emergency services
- Do not re-enter the building until it's declared safe

2. In Case of Fire:
- Evacuate immediately
- Call emergency services
- Use a fire extinguisher only if the fire is small and you're trained to do so

- Never try to extinguish a fire if you hear or smell gas escaping

3. If Someone is Overcome by Gas:
- Get them to fresh air immediately
- Call emergency services
- Begin CPR if the person is not breathing and you're trained to do so

Troubleshooting Tip: Conduct regular emergency drills with your family or coworkers to ensure everyone knows how to respond in a gas-related emergency.

Remember, working with gas lines is not just about technical skill – it's about maintaining a constant awareness of safety. By understanding and following these safety considerations and regulations, you're not just protecting yourself, but also your loved ones and your community.

In the next sections, we'll dive deeper into specific gas line installation and repair techniques, always keeping these crucial safety principles in mind. Are you ready to continue your journey towards becoming a responsible and knowledgeable gas line expert? Let's move forward, with safety as our guiding principle!

Chapter 2
Tools and Materials
Essential Tools for Gas Line Work

Imagine standing in front of your tool shed, about to embark on a gas line project. You reach for your tools, but suddenly you're struck with doubt. Are these the right tools for the job? Will they ensure your safety and the integrity of your work? Having the proper tools isn't just about convenience—it's a critical aspect of safe and effective gas line work.

When I first started working on gas lines, I made the mistake of thinking I could get by with my regular toolkit. I quickly learned that gas line work requires specialized tools designed for safety and precision. Let me walk you through the essential tools you'll need, explaining why each is important and how to use it effectively.

1. Pipe Wrenches

These heavy-duty wrenches are the workhorses of gas line installation and repair.

Key Features:
• Adjustable jaw for various pipe sizes
• Serrated teeth for a firm grip
• Available in different sizes (10", 14", 18" are most common for residential work)

Usage Tips:
• Always use two wrenches—one to hold the pipe steady, one to turn the fitting
• Protect pipe finishes by wrapping the pipe in a cloth before applying the wrench   **27**

Troubleshooting Tip: If a pipe wrench slips, clean the teeth with a wire brush. Dirty or worn teeth can lead to slippage and potential injury.

2. Tubing Cutters

Essential for making clean, straight cuts in copper or steel tubing.

Key Features:
- Rotating cutting wheel
- Adjustable for different tube diameters
- Some models designed specifically for tight spaces

Usage Tips:
- Tighten the cutter gradually as you rotate it around the tube
- Use a reamer to remove burrs after cutting

Troubleshooting Tip: If cuts are uneven, check the cutting wheel for wear. Replace the wheel or the entire cutter if necessary to ensure clean, safe cuts.

3. Thread Sealant and Tape

Crucial for creating leak-free connections in threaded fittings.

Types:
- Pipe dope (liquid or paste sealant)
- PTFE tape (also known as Teflon tape)

Usage Tips:
- Apply pipe dope in a thin, even layer on male threads only
- Wrap PTFE tape clockwise around male threads, overlapping by half the tape's width

Troubleshooting Tip: If you're experiencing leaks despite using sealant, you may be using too much. Excess sealant can prevent proper thread engagement. Clean the threads and reapply sparingly.

4. Gas Leak Detector

An indispensable safety tool for identifying gas leaks.

Types:
- Electronic detectors (most sensitive and versatile)
- Bubble solutions (simple but effective for visible connections)

Usage Tips:
- Test the detector before each use
- Move the sensor slowly along all connections and potential leak points

Troubleshooting Tip: If your electronic detector seems erratic, check for low batteries or sensor contamination. Clean the sensor according to manufacturer instructions and replace batteries regularly.

5. Pipe Threader

For creating precise threads on steel pipes.

Key Features:
- Dies for different pipe sizes
- Ratcheting handle for easier operation
- Some models are power-driven for larger jobs

Usage Tips:
- Use plenty of threading oil to reduce friction and improve thread quality
- Clean chips away frequently during threading

Troubleshooting Tip: If threads are coming out rough or incomplete, your dies may be dull. Replace them to ensure tight, leak-free connections.

6. Flaring Tool

Necessary for creating flared connections in soft copper tubing.

Key Features:
- Flaring bars to hold tubing
- Yoke and cone for shaping the flare

Usage Tips:
- Cut the tube squarely and remove all burrs before flaring
- Don't over-tighten the flaring cone, as this can lead to cracks

Troubleshooting Tip: If flares are coming out uneven or split, your flaring cone may be worn or damaged. Inspect it carefully and replace if necessary.

7. Pipe Bender

For creating smooth bends in copper or flexible gas lines.

Key Features:
- Different sizes for various pipe diameters
- Some models include measurement markings for precise bends

Usage Tips:
- Mark your desired bend location before starting
- Bend gradually to avoid kinking the pipe

Troubleshooting Tip: If you're getting kinks in your bends, you may be applying force too quickly. Try bending more gradually, and consider using a spring bender inside the pipe for additional support.

8. Reciprocating Saw with Metal Cutting Blade

Useful for cutting through existing pipes during renovations or repairs.

Usage Tips:
- Use a fine-toothed blade designed for metal cutting
- Secure the pipe well to prevent dangerous vibrations during cutting

Troubleshooting Tip: If the saw is binding or cutting slowly, your blade may be dull or you might be applying too much pressure. Let the saw do the work, and replace blades regularly.

9. Level and Measuring Tape

Essential for ensuring proper slope and accurate measurements.

Usage Tips:
- Use a 4-foot level for checking pipe slopes
- Measure twice, cut once to avoid wasting materials

Troubleshooting Tip: Regularly check your level for accuracy by flipping it end-for-end on a surface. If the bubble doesn't settle in the same spot, it's time for a replacement.

10. Personal Protective Equipment (PPE)

Safety should always come first in gas line work.

Essential PPE includes:
- Safety glasses or goggles
- Work gloves
- Steel-toed boots
- Hearing protection when using power tools

Usage Tips:
- Inspect all PPE before each use
- Replace any damaged or worn items immediately

Troubleshooting Tip: If your safety glasses are constantly fogging up, try an anti-fog spray or wipe. Clear vision is crucial for safe and accurate work.

11. Voltage Non-Contact Tester

While not directly used on gas lines, this tool is crucial for safety when working near electrical systems.

Usage Tips:
- Always test on a known live circuit before use
- Move slowly along wires and connections to detect voltage

Troubleshooting Tip: If the tester seems unreliable, replace the batteries. If issues persist, replace the entire unit – your safety is worth the investment.

Remember, having the right tools is just the beginning. Proper maintenance and correct usage of these tools are equally important. Always read and follow the manufacturer's instructions for each tool. Keep your tools clean and in good repair, and don't hesitate to replace them when they show signs of wear.

Investing in high-quality tools specifically designed for gas line work is an investment in your safety and the quality of your work. While it might seem tempting to cut corners or use makeshift solutions, remember that when it comes to gas lines, the stakes are too high for compromises.

In our next section, we'll explore the various materials used in gas line systems, helping you choose the right pipes, fittings, and components for your project. Are you ready to build your gas line toolkit and take your skills to the next level? Let's continue this journey towards mastering safe and effective gas line work!

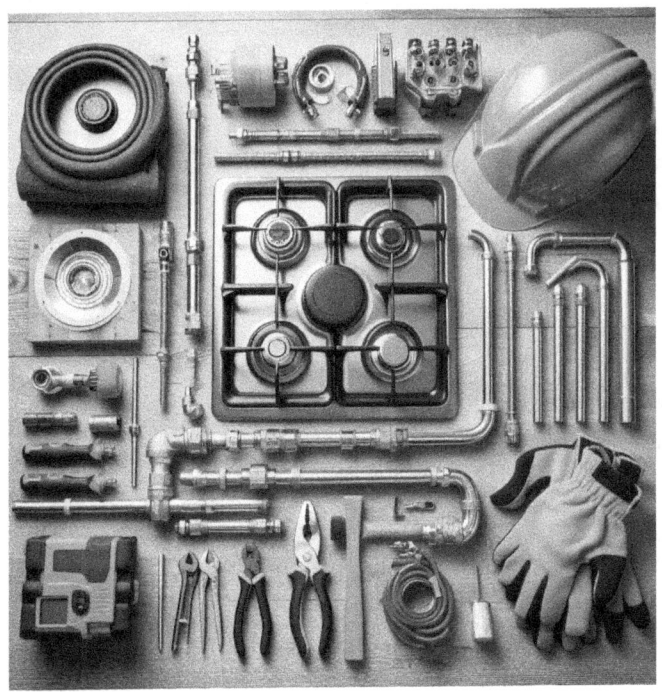

Pipe Materials and Fittings

Imagine you're standing in the plumbing aisle of your local hardware store, surrounded by a dizzying array of pipes and fittings. You know your gas line project depends on choosing the right materials, but how do you make sense of all these options? Understanding pipe materials and fittings is crucial for creating a safe, efficient, and long-lasting gas line system.

When I first faced this challenge, I felt overwhelmed by the choices. But as I learned about each material's properties and applications, I gained the confidence to make informed decisions. Let me guide you through the world of gas line materials and fittings, helping you choose the right components for your project.

1. Black Iron Pipe

Black iron pipe is the most common material used for gas lines in residential and commercial settings.

Key Features:
- Durable and long-lasting
- Resistant to high temperatures
- Available in various diameters and lengths

Applications:
- Main gas lines
- Branch lines to appliances
- Both indoor and outdoor use

Fittings:
- Threaded fittings (elbows, tees, couplings, etc.)
- Require pipe dope or PTFE tape for sealing

Troubleshooting Tip: If you notice rust on black iron pipes, it's usually just surface oxidation. However, if the rust is flaking or the pipe feels thin, it may be time for replacement. Regularly inspect visible pipes for signs of corrosion.

2. Corrugated Stainless Steel Tubing (CSST)

CSST is a flexible alternative to rigid piping, making it easier to install in tight spaces.

Key Features:
- Flexible and easy to route around obstacles
- Resistant to corrosion
- Requires fewer fittings than rigid pipe

Applications:
- Branch lines to appliances
- Retrofit installations where access is limited

Fittings:
- Special fittings designed for CSST
- Usually require special tools for installation

Troubleshooting Tip: CSST must be properly bonded to the building's electrical grounding system to prevent potential damage from lightning strikes. If you're unsure about the bonding, consult an electrician or CSST specialist.

3. Copper Tubing

While less common for gas lines, copper tubing is still used in some regions and applications.

Key Features:
- Corrosion-resistant
- Easy to work with (can be bent and shaped)
- Lightweight

Applications:
- Short runs to appliances
- Not approved for use in all jurisdictions - check local codes

Fittings:
- Flare fittings
- Compression fittings (in some cases)

Troubleshooting Tip: If you're using copper for gas lines, be sure to use tubing rated for fuel gas (marked with a "K" or "L"). Regular water pipe copper is not suitable for gas applications.

4. Polyethylene (PE) Pipe

PE pipe is commonly used for underground gas lines due to its durability and resistance to corrosion.

Key Features:
- Flexible and easy to install
- Highly resistant to corrosion and chemicals
- Long lifespan

Applications:
- Underground gas lines
- Not approved for above-ground or indoor use

Fittings:
- Special fittings designed for PE pipe
- Often require heat fusion for joining

Troubleshooting Tip: PE pipe can be damaged by UV radiation. If any portion of the pipe is exposed above ground, protect it with a UV-resistant coating or sleeve.

Understanding Fittings

Choosing the right fittings is just as important as selecting the proper pipe material. Here are some common types of fittings you'll encounter:

1. Threaded Fittings

Used primarily with black iron pipe.

Types:
- Elbows (45° and 90°)
- Tees
- Couplings
- Unions
- Caps and plugs

Usage Tips:
- Always use pipe dope or PTFE tape on male threads
- Tighten fittings securely, but avoid over-tightening which can crack the fitting

Troubleshooting Tip: If a threaded connection leaks, try disassembling it, cleaning the threads thoroughly, and reapplying sealant. If the leak persists, the threads may be damaged and the fitting should be replaced.

2. Flare Fittings

Common for connecting copper tubing and some types of CSST.

Key Components:
- Flare nut
- Flare sleeve (for some types)
- Flared end of the tube

Usage Tips:
- Ensure the flare is properly formed ($45°$ angle for gas lines)
- Avoid over-tightening, which can damage the flare

Troubleshooting Tip: If a flare fitting leaks, the flare may be improperly formed or damaged. Cut off the old flare, re-flare the tube, and reassemble the connection.

3. Compression Fittings

Used in some applications for connecting copper tubing.

Key Components:
- Compression nut
- Compression ring (ferrule)
- Fitting body

Usage Tips:
- Ensure the tube is cut square and free of burrs
- Tighten the nut until snug, then an additional 1/4 to 1/2 turn

Troubleshooting Tip: If a compression fitting leaks, try tightening it slightly more. If the leak persists, disassemble the fitting and check the ferrule for proper seating or damage. Replace the ferrule if necessary.

4. CSST Fittings

Specialized fittings designed for use with CSST.

Key Components:
- Fitting body
- Retainer ring
- Nut

Usage Tips:
- Follow manufacturer's instructions carefully
- Use the correct tools for cutting CSST and attaching fittings

Troubleshooting Tip: If a CSST fitting leaks, ensure it's properly tightened according to manufacturer specifications. Never reuse CSST fittings - always replace them if you need to disconnect and reconnect a line.

5. PE Pipe Fittings

Designed specifically for use with polyethylene pipe.

Types:
- Butt fusion fittings
- Electrofusion fittings
- Mechanical fittings (for transitions to other materials)

Usage Tips:
- Fusion fittings require special equipment and training
- Follow manufacturer's guidelines for proper installation

Troubleshooting Tip: If you suspect a leak in a fused PE joint, do not attempt to repair it yourself. These joints require specialized equipment and expertise to fix properly. Contact a professional PE pipe specialist.

Choosing the Right Materials

When selecting pipe materials and fittings for your gas line project, consider the following factors:

1. Local Codes and Regulations: Always check your local building codes to ensure the materials you choose are approved for use in your area.

2. Compatibility: Ensure all components are compatible with each other and with the type of gas you're using (natural gas or propane).

3. Pressure Rating: Verify that all materials are rated for the gas pressure in your system.

4. Environment: Consider factors like exposure to sunlight, soil conditions (for underground lines), and potential physical damage.

5. Installation Requirements: Some materials require special tools or skills to install properly. Ensure you have the necessary resources or consider hiring a professional.

6. Cost: While it's tempting to choose the cheapest option, remember that quality and safety should be your top priorities when it comes to gas lines.

Remember, working with gas lines is serious business. If you're ever in doubt about which materials to use or how to install them properly, don't hesitate to consult a professional. The safety of your home and family is worth the extra care and attention to detail.

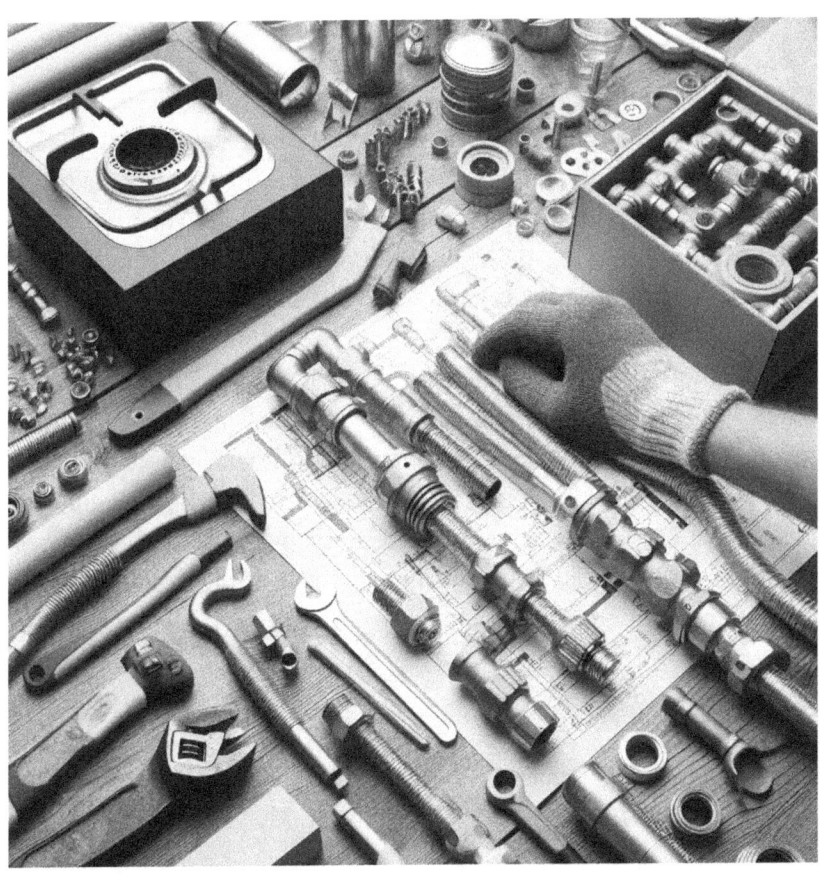

Safety Equipment

Picture this: You're about to start a gas line project in your home. You've got your tools ready and your materials selected, but there's a nagging feeling that you might be forgetting something crucial. That's where safety equipment comes in. When working with gas lines, having the right safety gear isn't just a precaution —it's an absolute necessity.

I remember my first gas line project. I was so focused on the technical aspects that I almost overlooked proper safety equipment. Thankfully, a more experienced friend pointed out this oversight. Now, I never start a gas-related task without first ensuring I have all the necessary safety gear. Let me walk you through the essential safety equipment you need, explaining why each item is important and how to use it effectively.

1. Gas Detector

A gas detector is your first line of defense against potentially deadly gas leaks.

Key Features:
- Detects natural gas, propane, and other combustible gases
- Audible and visual alarms
- Some models detect multiple types of gases

Usage Tips:
- Test the detector before each use
- Move the sensor slowly along all connections and potential leak points
- Pay attention to both the visual display and audible alerts

Troubleshooting Tip: If your gas detector seems unresponsive or gives erratic readings, check the batteries first. If replacing batteries doesn't solve the issue, the sensor may need cleaning or calibration. Follow the manufacturer's instructions for maintenance, or consider replacing the unit if it's old.

2. Personal Protective Equipment (PPE)

PPE is crucial for protecting yourself from various hazards associated with gas line work.

Essential PPE includes:

a) Safety Glasses or Goggles
Protect your eyes from debris, sparks, and potential gas spray.

Usage Tip: Ensure they fit comfortably and don't fog up easily. Consider anti-fog coatings or ventilated designs for clearer vision.

b) Work Gloves
Shield your hands from cuts, abrasions, and potential chemical exposure.

Usage Tip: Choose gloves that offer both protection and dexterity. Leather or heavy-duty nitrile gloves are good options for gas line work.

c) Steel-Toed Boots
Protect your feet from falling tools or materials.

Usage Tip: Ensure boots fit well and provide good ankle support to prevent trips and falls.

d) Flame-Resistant Clothing
Provides an extra layer of protection in case of flash fires.

Usage Tip: Look for clothing made from materials like Nomex or treated cotton. Ensure it fits properly without loose parts that could catch on tools or equipment.

Troubleshooting Tip: Regularly inspect all PPE for signs of wear or damage. Replace any compromised items immediately. Remember, PPE is only effective if it's in good condition and used correctly.

3. Fire Extinguisher

A must-have for any gas line work. Ensure you have the right type of extinguisher for gas fires.

Key Features:
- Class B or multi-purpose ABC rated
- Easily accessible in your work area
- Regularly inspected and maintained

Usage Tips:
- Familiarize yourself with the P.A.S.S. technique (Pull, Aim, Squeeze, Sweep)
- Keep the extinguisher within reach but not so close that you can't access it if a fire breaks out at your work site

Troubleshooting Tip: If your fire extinguisher's pressure gauge isn't in the green zone, or if it's past its expiration date, replace it immediately. A malfunctioning extinguisher can give a false sense of security.

4. Ventilation Equipment

Proper ventilation is crucial when working with gas to prevent the buildup of potentially explosive fumes.

Types:
- Portable fans
- Window fans
- Explosion-proof blowers for high-risk areas

Usage Tips:
- Position fans to create a cross-flow of air, pushing fumes out of the work area
- Ensure the air is being exhausted to a safe outdoor area, not just into another part of the building

Troubleshooting Tip: If you smell gas despite using ventilation, stop work immediately and reassess your ventilation setup. You may need more powerful fans or to reposition them for better airflow.

5. First Aid Kit

While prevention is key, being prepared for potential injuries is crucial.

Essential components:
- Burn treatment supplies
- Bandages and gauze
- Eyewash solution
- Antiseptic wipes

Usage Tip: Familiarize yourself with the contents of your first aid kit before starting work. Know how to use each item effectively.

Troubleshooting Tip: Regularly check your first aid kit and replace any used or expired items. Consider taking a first aid course to improve your ability to respond to emergencies.

6. Emergency Shut-Off Tool

A specialized tool for quickly shutting off the main gas valve in case of an emergency.

Key Features:
- Designed to fit standard gas meter valves
- Usually made of non-sparking materials

Usage Tip: Store this tool in an easily accessible location, and ensure all household members know where it is and how to use it.

Troubleshooting Tip: Periodically check that the tool fits your gas meter valve properly. Valves can sometimes change due to weathering or updates, requiring a different tool.

7. Communication Device

Always have a way to call for help in case of an emergency.

Options:
- Cell phone (ensure it's charged)
- Two-way radio for areas with poor cell reception

Usage Tip: Pre-program emergency numbers, including your gas company's emergency line, into your phone for quick access.

Troubleshooting Tip: If you're working in an area with poor cell reception, consider a signal booster or establish a check-in system with someone outside the work area.

8. Combustible Gas Signage

Proper warning signs alert others to the presence of gas and potential dangers.

Types:
- "No Smoking" signs
- "Combustible Gas" warning signs
- "Emergency Shut-Off" location markers

Usage Tip: Place signs prominently at entrances to work areas and near gas equipment.

Troubleshooting Tip: Regularly check that signs are visible and not obscured by equipment or debris. Replace any damaged or faded signs promptly.

9. Respirator

While proper ventilation should prevent the need for respiratory protection, a respirator can provide an extra layer of safety in certain situations.

Key Features:
- Rated for organic vapors
- Proper fit is crucial for effectiveness

Usage Tip: Undergo proper fit testing and training before relying on a respirator for protection.

Troubleshooting Tip: If you feel light-headed or smell gas while wearing a respirator, leave the area immediately. The respirator may not be fitting properly or may be overwhelmed by the concentration of gas.

Remember, safety equipment is only effective if it's in good condition, used correctly, and appropriate for the task at hand. Always inspect your safety gear before starting work, and never hesitate to replace or upgrade equipment if you have any doubts about its effectiveness.

Additionally, while having the right safety equipment is crucial, it's equally important to know how to use it effectively. Consider taking safety courses or training sessions specific to gas line work. Many local fire departments, community colleges, or trade schools offer such training.

Lastly, always remember that no piece of safety equipment can replace common sense and caution. If something feels unsafe, stop work immediately and reassess the situation. It's always better to err on the side of caution when working with gas lines.

Chapter 3
Planning Your Gas Line Installation
Assessing Your Needs

Picture this: You're standing in your kitchen, envisioning a beautiful new gas range, or perhaps you're in your backyard, dreaming of a cozy outdoor fire pit. But before you can bring these visions to life, you need to assess your gas line needs carefully. This crucial step sets the foundation for a safe, efficient, and code-compliant gas line installation.

When I first planned a gas line extension for my outdoor grill, I almost rushed into the project without a proper assessment. Thankfully, I took a step back and realized the importance of thorough planning. Let me guide you through the process of assessing your gas line needs, sharing insights I've gained from both successes and near-misses.

1. Identify Your Gas Appliances

Start by listing all the gas appliances you plan to install or connect. This might include:

- Kitchen appliances (stove, oven, cooktop)
- Heating systems (furnace, boiler, fireplace)
- Water heaters
- Clothes dryers
- Outdoor appliances (grill, fire pit, pool heater)

For each appliance, note down:
- Make and model
- BTU (British Thermal Unit) rating
- Required gas pressure
- Planned location in your home or property

Troubleshooting Tip: If you're unsure about an appliance's gas requirements, consult the manufacturer's specifications. Don't rely on guesswork – incorrect assumptions can lead to inadequate gas supply or safety issues.

2. Calculate Total Gas Load

Sum up the BTU ratings of all your appliances to determine your total gas load. This is crucial for ensuring your gas supply can handle all appliances running simultaneously.

Example calculation:
- Gas range: 65,000 BTU
- Water heater: 40,000 BTU
- Furnace: 80,000 BTU
- Total load: 185,000 BTU

Troubleshooting Tip: Always add a safety margin to your calculations. A good rule of thumb is to add 20% to your total load to account for future additions or higher-than-expected usage.

3. Assess Your Existing Gas Supply

If you're adding to an existing gas system, you need to determine if your current supply can handle the additional load.

Check:
- Meter capacity
- Existing pipe sizes
- Current gas pressure

You may need to consult with your gas company or a licensed plumber to get this information.

Troubleshooting Tip: If your existing supply seems borderline for your needs, it's better to upgrade now rather than risk inadequate supply later. Undersized gas lines can lead to poor appliance performance and potential safety hazards.

4. Map Out Your Gas Line Route

Sketch a rough layout of your property, marking:
- Existing gas meter location
- Planned locations for new appliances
- Potential routes for new gas lines

Consider:
- Shortest possible routes to minimize materials and potential leak points
- Accessibility for future maintenance
- Avoiding areas prone to physical damage (e.g., driveways, high-traffic areas)

Troubleshooting Tip: Watch out for potential obstacles like electrical wiring, water pipes, or load-bearing walls. You may need to adjust your route to avoid these, so it's best to identify them early in the planning stage.

5. Understand Local Codes and Regulations

Gas line installation is subject to strict regulations. Research local building codes, which may dictate:
- Approved materials for gas lines
- Required distances from electrical sources
- Ventilation requirements for gas appliances
- Necessary permits and inspections

Troubleshooting Tip: Don't assume that because something was allowed in a previous project, it's still compliant. Codes can change, so always check the most current regulations.

6. Consider Future Needs

Think ahead to potential future gas appliances or home additions. It's often more cost-effective to install a slightly larger gas line now than to upgrade later.

Questions to ask yourself:
- Might you want to add a gas fireplace in the future?
- Are you considering an outdoor kitchen down the line?
- Could you switch to a gas clothes dryer in the coming years?

Troubleshooting Tip: While planning for the future is wise, be careful not to oversize your system significantly. Oversized gas lines can lead to higher installation costs and potential safety issues.

7. Assess Your Skill Level and Resources

Be honest about your abilities and the time you can dedicate to this project.

Consider:
- Your experience with plumbing or gas line work
- Availability of necessary tools
- Time you can allocate to the project
- Budget for materials and potentially professional help

Troubleshooting Tip: If you're unsure about any aspect of the installation, it's better to consult a professional early in the planning stage. They can help refine your plan and may spot potential issues you've overlooked.

8. Plan for Safety

Incorporate safety measures into your plan from the start:
- Locations for emergency shut-off valves
- Proper ventilation for gas appliances
- Carbon monoxide detector placements
- Clear access to the gas meter

Troubleshooting Tip: Create an emergency plan as part of your assessment. Ensure all household members know how to shut off the gas in case of an emergency.

9. Document Everything

As you assess your needs, keep detailed notes and drawings. This documentation will be invaluable when:
- Applying for permits
- Consulting with professionals
- Purchasing materials
- Referring back during the installation process

Troubleshooting Tip: Take photos of your existing setup before planning changes. These can be helpful references and may be required for permit applications.

10. Get a Professional Assessment

Even if you plan to do the installation yourself, consider having a licensed gas fitter or plumber review your assessment. They can:
- Verify your load calculations
- Suggest optimizations to your plan
- Identify potential code issues
- Provide valuable insights from their experience

Troubleshooting Tip: If a professional suggests significant changes to your plan, don't be discouraged. View it as an opportunity to refine your assessment skills and create a safer, more efficient installation.

Remember, a thorough needs assessment is the foundation of a successful gas line installation. It helps ensure your system will be safe, efficient, and compliant with local regulations. Plus, it can save you time and money by identifying potential issues before you start the physical work.

Creating a Layout

Imagine you're an artist, and your canvas is your home's floor plan. The gas line layout you're about to create is like a intricate painting, where every line, curve, and connection point matters. A well-designed layout not only ensures efficient gas delivery but also contributes to the safety and longevity of your system.

When I first attempted to create a gas line layout, I underestimated its complexity. I quickly learned that a thoughtful, detailed layout is crucial for a successful installation. Let me guide you through the process of creating a comprehensive gas line layout, sharing insights I've gained from both triumphs and mistakes.

1. Start with a Detailed Floor Plan

Begin by creating or obtaining a detailed floor plan of your home or the area where you're installing gas lines. This should include:

- Accurate measurements of rooms and spaces
- Locations of existing gas lines and meter
- Positions of walls, doors, and windows
- Locations of other utility lines (electrical, water, sewer)

Tip: Use graph paper or a computer-aided design (CAD) program for precision. Each square can represent a specific measurement (e.g., 1 square = 1 foot).

Troubleshooting Tip: If you're working with an older home, don't assume existing plans are accurate. Always verify measurements yourself to avoid costly mistakes during installation.

Tip: Use a different color or line style for ventilation elements to distinguish them from gas lines.

2. Mark Appliance Locations

On your floor plan, clearly mark the locations of all gas appliances, both existing and planned. Include:

- Exact positions of appliances
- Required clearances around each appliance
- Gas input locations on the appliances

Tip: Use different colors or symbols for existing and planned appliances to easily distinguish between them.

Troubleshooting Tip: Double-check manufacturer specifications for each appliance. Required clearances or input locations may vary, even among similar appliances.

3. Plan Your Pipe Routes

Now, start drawing your gas line routes. Consider the following:

- Shortest possible path from the gas meter to each appliance
- Avoiding obstacles like electrical wiring, plumbing, and HVAC ducts
- Maintaining proper clearances from heat sources
- Accessibility for future maintenance and repairs
- Minimizing the number of fittings and turns (each adds resistance)

Tip: Use a different color or line style for your planned gas lines to distinguish them from existing lines and other elements on your plan.

Troubleshooting Tip: If you find yourself planning too many bends or a particularly long run, consider alternative routes. Each bend and extra length of pipe increases pressure drop in the system.

4. Indicate Pipe Sizes

Based on your gas load calculations from the needs assessment, determine and note the appropriate pipe sizes for each section of your layout. Remember:

- Pipe size may change as it branches off to different appliances
- Longer runs may require larger pipe diameters to maintain proper pressure

Tip: Label each section of pipe with its diameter and length for easy reference during installation and when creating your materials list.

Troubleshooting Tip: When in doubt, consult gas pipe sizing charts or use online calculators. Undersized pipes can lead to insufficient gas flow, while oversized pipes waste materials and can create other issues.

5. Mark Shut-Off Valves

Include shut-off valves in your layout. You should have:

- A main shut-off valve near where the gas line enters the building
- Individual shut-off valves for each appliance

Tip: Position valves in easily accessible locations, but not where they might be accidentally bumped or turned.

Troubleshooting Tip: If a valve location seems awkward or hard to reach in your layout, rethink it now. It's much easier to adjust on paper than during installation.

6. Plan for Future Expansion

If you anticipate adding gas appliances in the future, include provisions in your layout:

- Capped lines extending to potential future appliance locations
- Slightly oversized main lines to handle potential additional load

Tip: Mark these future provisions clearly on your plan so they're not overlooked during installation.

Troubleshooting Tip: While planning for the future is wise, be cautious about installing too many unused lines. Each additional line is a potential leak point and adds to your material costs.

7. Include Drip Legs

Plan for drip legs (also called sediment traps) in your layout. These should be installed:

- At the bottom of vertical pipes
- Before gas enters an appliance

Tip: Drip legs should be at least 3 inches long and capped at the bottom for easy cleaning.

Troubleshooting Tip: If you forget to include drip legs, you risk debris entering and damaging your appliances. Always double-check that they're included in your layout.

8. Note Pipe Support Locations

Indicate where you'll need pipe supports in your layout. Generally:

- Horizontal pipes need support every 6-8 feet
- Vertical pipes need support every floor or every 10 feet

Tip: Mark support locations with a distinct symbol on your layout.

Troubleshooting Tip: Insufficient support can lead to sagging pipes and potential leaks. If a run looks too long without support in your layout, add an additional support point.

9. Plan for Proper Sloping

Gas lines should have a slight slope to allow any condensation to drain towards the drip legs. In your layout, indicate:

- Direction of slope for each pipe run
- Degree of slope (typically 1/4 inch per 15 feet)

Tip: Use arrows to indicate slope direction on your layout.

Troubleshooting Tip: If achieving the proper slope seems difficult in any section of your layout, you may need to rethink your pipe route or consider adding additional drip legs.

10. Include Ventilation Details

If you're installing new gas appliances, include ventilation details in your layout:

- Locations of vents or flues
- Routes for exhaust pipes
- Required clearances around vents

Troubleshooting Tip: Inadequate ventilation can lead to dangerous carbon monoxide buildup. If you're unsure about ventilation requirements, consult a professional HVAC technician.

11. Add Dimensions and Notes

Enhance your layout with detailed dimensions and explanatory notes:

- Distances between key points
- Depths for buried lines
- Special instructions for complex sections

Tip: Use a consistent system for notes and dimensions to keep your layout clear and easy to read.

Troubleshooting Tip: What seems obvious to you now might not be clear during installation or to another person. When in doubt, add more details to your layout.

12. Review and Refine

Once you've completed your initial layout, take time to review and refine it:

- Check for any missed appliances or connections
- Ensure all pipe sizes are noted
- Verify that shut-off valves are properly placed
- Look for ways to simplify or optimize the layout

Tip: It can be helpful to have someone else review your layout. A fresh pair of eyes might spot issues or opportunities you've missed.

Troubleshooting Tip: If your layout feels overly complex, that's a red flag. Simplicity often leads to a safer, more efficient system. Don't hesitate to revise and simplify.

Remember, creating a detailed gas line layout is a crucial step in ensuring a safe, efficient, and code-compliant installation. It's your roadmap for the actual installation process and a valuable tool for obtaining permits and inspections.

Obtaining Permits and Inspections

Imagine you've meticulously planned your gas line installation, created a detailed layout, and you're eager to start. But wait! There's a crucial step that often trips up even experienced DIYers: obtaining the necessary permits and scheduling inspections. This process might seem like a bureaucratic hurdle, but it's actually a vital safeguard ensuring the safety and legality of your project.

When I first tackled a gas line installation, I was tempted to skip the permitting process to save time. Fortunately, a more experienced friend warned me about the potential consequences. Now, I understand that permits and inspections are not just legal requirements, but valuable safety checks that can prevent costly mistakes and dangerous situations.

Let's walk through the process of obtaining permits and scheduling inspections, with insights I've gained from navigating this system multiple times.

1. Research Local Requirements

Every jurisdiction has its own specific requirements for gas line installations. Start by researching:

- Your local building department's website
- City or county code enforcement office
- State-level regulations

Look for information on:
- Types of permits required for gas line work
- Who can apply for permits (homeowner vs. licensed contractor)
- Specific code requirements for gas installations

Tip: Many jurisdictions now offer online resources and even permit applications. Check if your area provides these convenient options.

Troubleshooting Tip: If information seems unclear or contradictory, don't hesitate to call the permit office directly. It's better to ask questions early than to make assumptions that could cause problems later.

2. Prepare Your Application Materials

Typically, you'll need to submit:

- Completed permit application form
- Detailed plan of your gas line installation
- List of materials to be used
- Copy of your property survey or plot plan
- Proof of homeowner's insurance
- Contractor's license information (if applicable)

Tip: Create a checklist of required documents to ensure you don't miss anything.

Troubleshooting Tip: If you're unsure about any part of the application, ask for clarification before submitting. Incomplete or incorrect applications can significantly delay your project.

3. Submit Your Application

Once you've gathered all necessary materials:

- Submit your application in person or online (if available)
- Pay any required fees
- Get a receipt and expected timeline for review

Tip: Keep copies of everything you submit, including proof of payment.

Troubleshooting Tip: If you haven't heard back within the expected timeline, follow up politely. Sometimes applications can get overlooked in busy offices.

4. Review Process

The building department will review your application to ensure it meets all local codes and regulations. They may:

- Approve your application as-is
- Request additional information or clarification
- Suggest modifications to your plan

Tip: Respond promptly to any requests for additional information to keep your application moving forward.

Troubleshooting Tip: If your application is denied, don't get discouraged. Ask for a detailed explanation of why it was rejected and what changes are needed for approval.

5. Obtain Your Permit

Once approved, you'll receive your permit. This document:

- Gives you legal permission to proceed with your installation
- Outlines any specific conditions or requirements
- Specifies required inspections

Tip: Make multiple copies of your permit. Keep one with your project documents and post one visibly at the work site.

Troubleshooting Tip: Read through your permit carefully. There may be specific requirements or conditions that weren't obvious during the application process.

6. Schedule Inspections

Most gas line installations require multiple inspections:

- Rough-in inspection (before walls are closed)
- Pressure test inspection
- Final inspection

To schedule:
- Contact the building department or use their online scheduling system
- Provide your permit number and requested inspection type
- Get a confirmed date and time slot

Tip: Schedule inspections well in advance, as popular time slots can fill up quickly.

Troubleshooting Tip: If you're not sure you'll be ready by the inspection date, it's better to reschedule in advance rather than fail an inspection for incomplete work.

7. Prepare for Inspections

Before each inspection:

- Ensure the relevant work is complete and accessible
- Have your approved plans and permit on site
- Prepare any required testing equipment (e.g., pressure gauge for leak tests)
- Clear the area of any obstacles that might impede the inspector's work

Tip: Do a self-inspection using your local code requirements as a checklist before the official inspection.

Troubleshooting Tip: If you're unsure about any aspect of the inspection process, call the building department and ask. Inspectors appreciate when homeowners are prepared and knowledgeable.

8. During the Inspection

When the inspector arrives:

- Provide access to all areas of the installation
- Be prepared to answer questions about your work
- Take notes on any issues or recommendations the inspector mentions

Tip: Treat the inspector as a valuable resource. They can often provide insights that will improve your installation or future projects.

Troubleshooting Tip: If the inspector identifies issues, don't argue. Ask for clarification on what needs to be corrected and how to bring it up to code.

9. After the Inspection

The inspector will provide one of three outcomes:

- Pass: You're clear to move to the next phase or finish your project
- Conditional Pass: Minor issues need correction, but you can proceed
- Fail: Significant issues need to be addressed before you can continue

Tip: If you pass, keep the inspection report with your project documents. You may need it for future reference or when selling your home.

Troubleshooting Tip: If you fail an inspection, don't panic. Use it as a learning opportunity. Schedule a re-inspection as soon as you've addressed the issues.

10. Final Approval

After passing all required inspections:

- Ensure you receive final approval documentation
- File this with your other project documents
- Consider informing your homeowner's insurance of the completed, approved installation

Tip: Take photos of your completed, approved installation for your records.

Troubleshooting Tip: If there's any delay in receiving final approval documentation, follow up with the building department. You want official proof that your installation is fully approved and code-compliant.

Remember, while the permit and inspection process might seem tedious, it's designed to ensure your safety and the integrity of your home. It can also protect you legally and financially. An approved, inspected installation can increase your home's value and prevent issues with insurance claims or future sales.

Chapter 4
Gas Line Installation Techniques
Measuring and Cutting Pipes

Picture yourself standing in your workspace, shiny new pipes laid out before you, ready to begin your gas line installation. The first crucial step? Measuring and cutting those pipes to the perfect length. It might seem simple, but precision here is key to a safe, efficient, and leak-free gas system.

I remember my first time measuring and cutting pipes for a gas line. I was so focused on getting the length right that I nearly forgot about other crucial factors like accounting for fittings and proper tool use. Let me walk you through the process, sharing the insights I've gained to help you avoid common pitfalls and achieve professional-quality results.

1. Gathering Your Tools

Before you begin, ensure you have the right tools for the job:

- Tape measure
- Pipe cutter (for copper or soft steel)
- Hacksaw (for black iron pipe)
- Deburring tool
- Marker or pencil
- Safety glasses
- Work gloves

Tip: Invest in quality tools. A good pipe cutter or hacksaw will make cleaner cuts and last longer.

Troubleshooting Tip: If your cuts are rough or jagged, your cutting tool might be dull. Replace the blade or cutting wheel for better results.

2. Measuring Accurately

Precise measurement is crucial for a proper fit. Here's how to do it right:

a) Measure twice, cut once:
- Take your initial measurement
- Double-check it before marking
- Consider measuring from multiple reference points to ensure accuracy

b) Account for fittings:
- Remember that pipe fittings will add length to your run
- For threaded fittings, add about 3/4 inch to your measurement for each fitting
- For compression or flare fittings, add the depth of the fitting

c) Use a consistent measuring technique:
- Always measure from the same point on the pipe (e.g., end to end or center to center)

Tip: Use a small piece of masking tape to mark your cut point. It's easier to see and won't rub off like pencil marks can.

Troubleshooting Tip: If you consistently end up with pipes that are slightly too long or short, you might be forgetting to account for the width of your cutting tool. Adjust your marking point slightly to compensate.

3. Cutting Different Types of Pipe

Different pipe materials require different cutting techniques:

a) Copper Pipe:
- Use a tube cutter for clean, straight cuts
- Tighten the cutter gradually as you rotate it around the pipe
- Continue until the pipe is completely cut through

b) Black Iron Pipe:
- Use a hacksaw with a fine-toothed blade (at least 32 teeth per inch)
- Secure the pipe in a vice to prevent movement
- Cut slowly and steadily, applying even pressure

c) Flexible Gas Lines (CSST):
- Use specialized CSST cutters
- Ensure a clean, square cut to maintain the integrity of the line

Tip: For all types of pipe, mark a cutting line all the way around the pipe to help you keep your cut straight.

Troubleshooting Tip: If you're having trouble getting a straight cut on black iron pipe, try wrapping a piece of paper around the pipe and aligning the edges. Use this as a guide for your hacksaw.

4. Deburring and Cleaning

After cutting, it's crucial to clean up the cut end:

a) Deburring:
- Use a deburring tool or file to remove any sharp edges or burrs
- For copper pipe, use the reaming blade on your tube cutter

b) Cleaning:
- Wipe the pipe end clean with a rag
- For copper pipe, use fine sandpaper or steel wool to clean the outside of the pipe end

Tip: Always deburr both the inside and outside of the pipe. Sharp edges can damage gaskets in fittings or create turbulence in gas flow.

Troubleshooting Tip: If you're having trouble getting a good seal with fittings, double-check that you've properly deburred the pipe ends. Even small burrs can prevent a proper fit.

5. Checking Your Work

Before moving on to the next step:

- Measure the cut pipe again to ensure it's the correct length
- Check that the cut is square (at a 90-degree angle to the pipe length)
- Ensure all burrs and debris have been removed

Tip: Use a small carpenter's square to check if your cut is truly square.

Troubleshooting Tip: If your cut isn't square, you may need to trim a small amount off the end to correct it. It's better to have a slightly shorter, square-cut pipe than one that's the right length but cut at an angle.

6. Special Considerations for Gas Lines

When cutting pipes for gas lines, keep these additional points in mind:

- Cleanliness is crucial. Any debris in the pipe can cause issues with gas flow or damage appliances.
- Be extra cautious about creating sparks when cutting, especially if you're working near existing gas lines.
- If you're cutting CSST, be careful not to damage the corrugations. They're crucial for the pipe's flexibility and strength.

Tip: If you're cutting pipe that will be visible (e.g., for a gas lamp), take extra care to make your cuts as clean and straight as possible for aesthetic reasons.

Troubleshooting Tip: If you accidentally cut a pipe too short, don't try to "stretch" it to fit. It's safer and more effective to cut a new piece of the correct length.

7. Proper Handling and Storage

After cutting:

- Cap or tape the ends of cut pipes to prevent debris from entering
- Store pipes in a clean, dry area to prevent corrosion or contamination
- Keep different types and sizes of pipe separated to avoid confusion

Tip: Label cut pieces with their intended location in your gas line system. This can save time and prevent mix-ups during installation.

Troubleshooting Tip: If you notice any rust or corrosion on stored pipes, clean them thoroughly before use. If the corrosion is severe, it's safer to use a new piece of pipe.

Remember, precise measuring and clean cutting are the foundation of a safe and efficient gas line installation. Take your time with these steps, and don't hesitate to re-cut if something isn't quite right. It's always better to use a little extra material than to compromise on the quality and safety of your installation.

Threading and Joining Pipes

Imagine you've just finished cutting your pipes to the perfect length. Now comes a critical step that can make or break your gas line installation: threading and joining those pipes. This process is where your gas line truly comes together, creating a secure, leak-free system that will safely deliver gas to your appliances.

I vividly remember my first attempt at threading pipes. I was so focused on getting the threads cut that I nearly forgot about proper cleaning and sealing. Over time, I've learned that each step in this process is crucial for a safe and effective installation. Let me guide you through the intricacies of threading and joining pipes, sharing the wisdom I've gained to help you achieve professional-quality results.

1. Threading Pipes

Threading creates the spiral grooves that allow pipes to be securely joined with fittings. Here's how to do it right:

Tools You'll Need:
- Pipe threader (manual or power)
- Threading oil
- Pipe vice
- Wire brush
- Clean rags

Steps:
a) Secure the pipe:
- Clamp the pipe firmly in a pipe vice
- Ensure at least 6 inches of pipe extend beyond the vice for threading

b) Apply threading oil:
- Liberally coat the end of the pipe with threading oil
- This lubricates the die and helps create clean threads

c) Position the threader:
- Place the threader squarely on the end of the pipe
- Ensure it's straight to create even threads

d) Start threading:
- Apply firm, steady pressure as you turn the threader clockwise
- Continue until you've threaded the desired length (usually about 3/4 inch for most fittings)

e) Clean the threads:
- Use a wire brush to remove any metal shavings
- Wipe the threads clean with a rag

Tip: Practice on some scrap pieces of pipe before threading pipes for your actual installation. This will help you get a feel for the process.

Troubleshooting Tip: If your threads are coming out rough or uneven, you might be applying too much pressure or using dull dies. Try easing up on the pressure and check your threading tool for wear.

2. Joining Threaded Pipes

Once you have threaded pipes, it's time to join them. Here's the process:

Tools and Materials:
- Pipe wrench
- Thread sealant (pipe dope or PTFE tape)
- Clean rags

Steps:

a) Clean the threads:
- Ensure both the male and female threads are clean and free of debris

b) Apply thread sealant:
- For pipe dope: Apply a thin, even layer to the male threads
- For PTFE tape: Wrap 3-4 turns around the male threads in a clockwise direction

c) Join the pipes:
- Hand-tighten the fitting onto the pipe
- Use a pipe wrench to tighten an additional 1-2 turns
- Be careful not to over-tighten, which can damage the threads or fitting

Tip: When using two pipe wrenches (one to hold the pipe, one to turn the fitting), wrap the pipe with a rag first to prevent damage to the finish.

Troubleshooting Tip: If you're having trouble getting a leak-free seal, disassemble the joint, clean the threads thoroughly, and start over with fresh sealant. Avoid the temptation to simply add more sealant to a leaky joint.

3. Joining Copper Pipes

For copper pipes, soldering (also called sweating) is the most common joining method:

Tools and Materials:
- Propane torch
- Lead-free solder
- Flux and flux brush
- Emery cloth or sand paper

- Clean rags

Steps:

a) Clean the pipe and fitting:
- Use emery cloth to clean the outside of the pipe end and the inside of the fitting
- The surfaces should be bright and shiny

b) Apply flux:
- Use a flux brush to apply a thin, even layer of flux to both surfaces
- This helps the solder flow and adhere

c) Assemble the joint:
- Push the pipe fully into the fitting
- Twist slightly to spread the flux

d) Heat the joint:
- Use the torch to heat the fitting evenly
- When the flux begins to bubble, it's ready for solder

e) Apply solder:
- Touch the solder to the joint opposite the flame
- When the joint is hot enough, the solder will melt and be drawn into the joint

f) Cool and clean:
- Allow the joint to cool naturally
- Wipe away any excess flux with a damp rag

Tip: Keep a fire extinguisher nearby when soldering, and be aware of your surroundings to prevent accidental fires.

Troubleshooting Tip: If solder isn't flowing into the joint, the pipe may not be hot enough, or there might not be enough flux. Remove the heat, let it cool, clean the joint, and start over.

4. Joining CSST (Corrugated Stainless Steel Tubing)

CSST requires special fittings and techniques:

Tools and Materials:
- CSST fitting kit
- CSST cutter
- Torque wrench

Steps:
a) Cut the CSST:
- Use a specialized CSST cutter to ensure a clean, square cut

b) Slide on compression nut and sleeve:
- These come with the CSST fitting kit

c) Insert the fitting:
- Push the fitting into the CSST until it stops

d) Tighten the nut:
- Use a torque wrench to tighten the nut to the manufacturer's specifications

Tip: Always follow the specific instructions provided with your CSST system, as different brands may have slightly different procedures.

Troubleshooting Tip: If you're having trouble getting the fitting to seal, double-check that you've cut the CSST squarely and that there's no damage to the corrugations near the cut end.

5. Testing Joints

After joining pipes, it's crucial to test for leaks:

- Apply a soap solution to all joints
- Pressurize the system (usually with air or nitrogen)
- Look for bubbles, which indicate leaks

Tip: Make your own leak detection solution by mixing a tablespoon of dish soap with a cup of water.

Troubleshooting Tip: If you find a leak, mark its location, depressurize the system, and repair the joint before retesting.

Remember, proper threading and joining of pipes is critical for the safety and efficiency of your gas line system. Take your time, follow these steps carefully, and don't hesitate to seek professional help if you're unsure about any part of the process. It's always better to be safe than sorry when dealing with gas lines.

Installing Shut-off Valves

Imagine this scenario: You're in the middle of cooking dinner when you smell gas. Your heart races as you realize you need to shut off the gas supply immediately. This is where properly installed shut-off valves become your best friend. They're not just a safety feature; they're a crucial component that can prevent a minor issue from becoming a major disaster.

When I first installed shut-off valves, I underestimated their importance and the precision required for proper installation. Over time, I've learned that these small components play a big role in the safety and functionality of your gas line system. Let me guide you through the process of installing shut-off valves, sharing insights I've gained to help you create a safer, more manageable gas system.

1. Understanding Shut-off Valves

Before we dive into installation, let's understand the types of shut-off valves:

a) Ball Valves:
- Most common for gas lines
- Quarter-turn operation
- Provide a visual indication of open/closed status

b) Gate Valves:
- Less common in modern installations
- Require multiple turns to open/close
- Can be prone to sticking if not used regularly

c) Safety Shut-off Valves:
- Automatically close if they detect a gas leak or fire
- Often required by code in commercial installations

Tip: For most residential applications, ball valves are the preferred choice due to their reliability and ease of use.

Troubleshooting Tip: If you're replacing an old gate valve with a ball valve, ensure the new valve is rated for gas use and meets local code requirements.

2. Choosing the Right Valve

Selecting the appropriate valve is crucial:

- Ensure the valve is specifically rated for natural gas or propane (as applicable)
- Match the valve size to your pipe size
- Check that the valve's pressure rating exceeds your system's maximum pressure
- Verify that the valve meets local code requirements

Tip: Look for valves with a clear "open" and "closed" indicator to avoid confusion in emergencies.

Troubleshooting Tip: If you're unsure about valve selection, consult with a professional or your local building department. Using the wrong type of valve can be dangerous and may violate building codes.

3. Determining Valve Locations

Proper placement of shut-off valves is key to a safe and convenient system:

a) Main Shut-off Valve:
- Install near where the gas line enters the building
- Ensure it's easily accessible in emergencies

b) Appliance Shut-off Valves:
- Install within 6 feet of each gas appliance
- Position for easy access, but not where they could be accidentally bumped or turned

c) Additional Valves:
- Consider installing valves at branch points in your gas line system
- Add valves to isolate sections for future maintenance or modifications

Tip: Create a map of your gas system showing all valve locations and keep it in an easily accessible place.

Troubleshooting Tip: If you find it difficult to place a valve in an easily accessible location, consider rerouting the pipe. Accessibility is crucial for safety and maintenance.

4. Installing the Valve

Now, let's go through the installation process:

Tools and Materials:
- Pipe wrenches
- Thread sealant (pipe dope or PTFE tape)
- Shut-off valve
- Pipe sections (if needed to accommodate the valve)

Steps:
a) Prepare the area:
- Ensure the gas is shut off at the meter
- Ventilate the area well

b) Cut the pipe (if necessary):
- Measure and mark where the valve will be installed
- Cut the pipe using appropriate tools (as discussed in section 4.1)

c) Clean and prepare the threads:
- Use a wire brush to clean the pipe threads
- Apply thread sealant to the male threads

d) Install the valve:
- Thread the valve onto the pipe, ensuring it's oriented correctly for gas flow
- Tighten with a pipe wrench, but be careful not to over-tighten

e) Connect the downstream pipe:
- Apply thread sealant to the valve's outlet threads
- Thread on the downstream pipe and tighten

f) Check alignment:
- Ensure the valve handle can fully open and close without obstruction

Tip: When tightening the valve, use two wrenches - one to hold the valve body steady and one to tighten the pipe. This prevents damage to the valve.

Troubleshooting Tip: If the valve doesn't align properly after tightening, don't try to force it. Instead, loosen the connections and adjust the pipe alignment before retightening.

5. Testing the Installation

After installation, it's crucial to test for leaks:

a) Pressurize the system:
- Usually done with air or nitrogen to a pressure specified by local codes

b) Apply leak detection solution:
- Use a commercial solution or mix dish soap with water
- Apply to all joints and the valve body

c) Check for bubbles:
- Bubbles indicate a leak
- Pay special attention to the valve stem and bonnet

d) Test valve operation:
- Open and close the valve several times to ensure smooth operation

Tip: Allow the test pressure to hold for at least 15 minutes to detect even small leaks.

Troubleshooting Tip: If you detect a leak, don't try to tighten the joint further while the system is pressurized. Release the pressure, disassemble the joint, clean it, apply new sealant, and reassemble before retesting.

6. Labeling and Documentation

After successful installation:

- Label each valve clearly with its purpose (e.g., "Main Gas Shut-off", "Kitchen Stove Shut-off")

- Update your gas system map with the new valve location
- Consider attaching a tag to rarely used valves with instructions for operation

Tip: Use durable, weather-resistant labels that won't fade or fall off over time.

Troubleshooting Tip: If you find old, unclear, or missing labels during your installation, take the time to replace or add labels to all valves in your system.

7. Maintenance

To ensure long-term functionality:

- Operate each valve at least once a year to prevent sticking
- Check for signs of corrosion or damage during annual inspections
- Replace any valves that show signs of wear or damage

Tip: Mark your calendar for annual valve checks to ensure you don't forget this important maintenance task.

Troubleshooting Tip: If a valve becomes difficult to turn, don't force it. This could cause damage or breakage. Instead, have it inspected and replaced if necessary.

Remember, properly installed shut-off valves are your first line of defense against gas leaks and a key component in maintaining a safe gas system. Take your time with installation, double-check your work, and never hesitate to call a professional if you're unsure about any aspect of the process.

Troubleshooting Tip: If you're testing a system with existing appliances connected, consult the appliance manufacturers' guidelines. Some appliances may need to be disconnected or bypassed during high-pressure testing.

4. Connecting the Test Equipment

Now, let's set up for the test:

a) Connect your pressure gauge to the system, usually at the furthest point from where you'll introduce the test pressure
b) Attach your air compressor or nitrogen tank to the system, typically near the main shut-off valve
c) Ensure all connections are tight and secure

Tip: Use appropriate adapters and fittings to ensure a secure connection between your test equipment and the gas system.

Troubleshooting Tip: If you're having trouble getting a tight seal for your test equipment, try using PTFE tape on the connections. Just remember to remove it after testing if it's not rated for gas use.

5. Pressurizing the System

Time to build up the pressure:

a) Slowly open the valve on your pressure source (compressor or nitrogen tank)
b) Watch the pressure gauge and stop when you reach the required test pressure
- This is typically 1.5 times the working pressure, often around 15 PSI for residential systems, but check your local codes for specific requirements

Pressure Testing

Imagine you've just completed your gas line installation. Everything looks perfect, but there's one crucial question remaining: Is it truly safe and leak-free? This is where pressure testing comes in. It's the final checkpoint, the ultimate test that ensures your hard work will result in a safe and efficient gas system.

When I first conducted a pressure test, I was nervous. The stakes felt high, and I wasn't sure what to expect. But I quickly learned that with the right approach, pressure testing is not just a safety measure—it's a confidence builder. Let me walk you through the process, sharing the insights I've gained to help you conduct a thorough and effective pressure test.

1. Understanding Pressure Testing

Pressure testing involves pressurizing your gas line system and monitoring it for any pressure drops, which would indicate a leak. Here's why it's crucial:

- Ensures all connections are tight and leak-free
- Verifies the integrity of the entire system
- Required by most building codes before a system can be put into service

Tip: Think of pressure testing as your system's final exam. It's your chance to catch and fix any issues before gas is introduced.

Troubleshooting Tip: If you're feeling overwhelmed, remember that a successful pressure test is your best assurance of a job well done. Take your time and don't rush the process.

2. Gathering Your Equipment

You'll need the following for a proper pressure test:

- Pressure gauge (capable of reading up to 15 PSI residential systems)
- Air compressor or nitrogen tank
- Appropriate fittings to connect the pressure source system
- Leak detection solution (commercial product or dis mixed with water)
- Stopwatch or timer
- Safety glasses and gloves

Tip: Invest in a quality pressure gauge. An accurate gai crucial for a reliable test.

Troubleshooting Tip: If you're unsure about your ga accuracy, consider having it calibrated or testing it again known accurate gauge before use.

3. Preparing Your System

Before you begin the test:

a) Ensure all pipes are connected and valves are installed
b) Cap off any open ends of the system
c) Verify that all appliance shut-off valves are closed
d) Double-check that the main gas valve is closed and the syste is disconnected from the gas supply

Tip: Use a checklist to ensure you haven't missed any open end or connections.

Pressure Testing

Imagine you've just completed your gas line installation. Everything looks perfect, but there's one crucial question remaining: Is it truly safe and leak-free? This is where pressure testing comes in. It's the final checkpoint, the ultimate test that ensures your hard work will result in a safe and efficient gas system.

When I first conducted a pressure test, I was nervous. The stakes felt high, and I wasn't sure what to expect. But I quickly learned that with the right approach, pressure testing is not just a safety measure—it's a confidence builder. Let me walk you through the process, sharing the insights I've gained to help you conduct a thorough and effective pressure test.

1. Understanding Pressure Testing

Pressure testing involves pressurizing your gas line system and monitoring it for any pressure drops, which would indicate a leak. Here's why it's crucial:

- Ensures all connections are tight and leak-free
- Verifies the integrity of the entire system
- Required by most building codes before a system can be put into service

Tip: Think of pressure testing as your system's final exam. It's your chance to catch and fix any issues before gas is introduced.

Troubleshooting Tip: If you're feeling overwhelmed, remember that a successful pressure test is your best assurance of a job well done. Take your time and don't rush the process.

2. Gathering Your Equipment

You'll need the following for a proper pressure test:

- Pressure gauge (capable of reading up to 15 PSI for most residential systems)
- Air compressor or nitrogen tank
- Appropriate fittings to connect the pressure source to your system
- Leak detection solution (commercial product or dish soap mixed with water)
- Stopwatch or timer
- Safety glasses and gloves

Tip: Invest in a quality pressure gauge. An accurate gauge is crucial for a reliable test.

Troubleshooting Tip: If you're unsure about your gauge's accuracy, consider having it calibrated or testing it against a known accurate gauge before use.

3. Preparing Your System

Before you begin the test:

a) Ensure all pipes are connected and valves are installed
b) Cap off any open ends of the system
c) Verify that all appliance shut-off valves are closed
d) Double-check that the main gas valve is closed and the system is disconnected from the gas supply

Tip: Use a checklist to ensure you haven't missed any open ends or connections.

Troubleshooting Tip: If you're testing a system with existing appliances connected, consult the appliance manufacturers' guidelines. Some appliances may need to be disconnected or bypassed during high-pressure testing.

4. Connecting the Test Equipment

Now, let's set up for the test:

a) Connect your pressure gauge to the system, usually at the furthest point from where you'll introduce the test pressure
b) Attach your air compressor or nitrogen tank to the system, typically near the main shut-off valve
c) Ensure all connections are tight and secure

Tip: Use appropriate adapters and fittings to ensure a secure connection between your test equipment and the gas system.

Troubleshooting Tip: If you're having trouble getting a tight seal for your test equipment, try using PTFE tape on the connections. Just remember to remove it after testing if it's not rated for gas use.

5. Pressurizing the System

Time to build up the pressure:

a) Slowly open the valve on your pressure source (compressor or nitrogen tank)
b) Watch the pressure gauge and stop when you reach the required test pressure
- This is typically 1.5 times the working pressure, often around 15 PSI for residential systems, but check your local codes for specific requirements

c) Close the valve on your pressure source

Tip: Pressurize the system slowly. A rapid pressure increase can damage gauges and potentially the system itself.

Troubleshooting Tip: If you can't reach the required pressure, you likely have a large leak. Perform a preliminary leak check with soapy water before continuing.

6. Conducting the Test

Now for the actual test:

a) Once at test pressure, let the system stabilize for about 10 minutes
b) Record the initial pressure and start your timer
c) Monitor the pressure for the required test duration (often 15 minutes to 24 hours, depending on local codes)
d) Record the final pressure at the end of the test period

Tip: During longer tests, account for temperature changes which can affect pressure readings. Try to keep the system at a consistent temperature.

Troubleshooting Tip: If you notice a significant pressure drop immediately, it's likely you have a substantial leak. Depressurize the system and perform a thorough leak check before retesting.

7. Interpreting the Results

After the test period:

- No pressure drop: Congratulations! Your system is likely leak-free.

- Slight pressure drop: Small leaks may be present. Further investigation is needed.
- Significant pressure drop: Large leaks are present. A thorough inspection is required.

Tip: Even a small pressure drop can indicate a leak. Don't ignore it just because it seems minor.

Troubleshooting Tip: If you detect a pressure drop, don't immediately assume it's a leak. Check for temperature changes or other factors that could affect pressure before conducting a leak search.

8. Leak Detection

If your test indicates a possible leak:

a) Mix a solution of dish soap and water or use a commercial leak detection fluid
b) Apply this solution to all joints, connections, and valves
c) Look for bubbles forming, which indicate a leak
d) Mark any leaks you find for later repair

Tip: Pay extra attention to areas that were difficult to access or install. These are often prime spots for leaks.

Troubleshooting Tip: No bubbles but still losing pressure? Consider using an electronic gas detector for more sensitive leak detection.

9. Retesting After Repairs

If you find and repair leaks:

a) Depressurize the system completely
b) Make necessary repairs
c) Repeat the entire pressure test process

Tip: Always retest the entire system, not just the repaired areas. A repair in one area can sometimes cause issues elsewhere.

Troubleshooting Tip: If you're repeatedly failing pressure tests, consider bringing in a professional to inspect your work. Fresh eyes can often spot issues you might have overlooked.

10. Documentation

After a successful test:

- Record the test pressure, duration, and results
- Note any repairs made
- Take photos of your gauge readings if possible
- Keep this documentation for future reference and inspections

Tip: Create a detailed report of your pressure test. This can be invaluable for future maintenance or if you ever sell your home.

Troubleshooting Tip: If your documentation is questioned during an inspection, offer to conduct another test in the inspector's presence.

Remember, a successful pressure test is your assurance that you've created a safe, leak-free gas system. It's the final stamp of approval on your hard work. Take pride in completing this crucial step, but always prioritize safety. If at any point you feel unsure or uncomfortable with the process, don't hesitate to call in a professional.

Chapter 5
Connecting Gas Appliances
Stove Installation

Picture this: You've just brought home a beautiful new gas stove, the centerpiece of your kitchen renovation. It promises delicious meals and precise temperature control, but first, you need to connect it safely to your gas line. This final step can feel daunting, but with the right approach, it's a task you can master.

I remember my first gas stove installation. I was excited but also nervous about working with gas connections. Over time, I've learned that careful preparation and attention to detail make all the difference. Let me guide you through the process of installing a gas stove, sharing the insights I've gained to help you complete this task safely and confidently.

1. Safety First

Before we begin, let's review some crucial safety measures:

- Ensure proper ventilation in the work area
- Have a fire extinguisher nearby
- Know where your main gas shut-off valve is located
- Never smoke or have open flames in the area while working

Tip: If at any point you smell gas or feel unsure, stop immediately and call a professional.

Troubleshooting Tip: If you've recently moved into a new home and are unsure about your gas system, consider having a professional inspection before attempting any installations.

2. Gather Your Tools and Materials

You'll need:

- Adjustable wrench
- Pipe wrench
- Gas leak detector or soapy water solution
- Flexible gas connector (if allowed by local codes)
- Pipe joint compound (gas-rated)
- Teflon tape (gas-rated)
- Shut-off valve (if not already installed)

Tip: Always use tools and materials specifically rated for gas line work.

Troubleshooting Tip: If your local code doesn't allow flexible connectors, you'll need to use rigid pipe. This may require additional tools and skills, so consider professional help if you're not comfortable with this.

3. Prepare the Area

Before bringing in the stove:

a) Measure the space to ensure the stove will fit
b) Clean the area thoroughly
c) Locate the gas line stub-out in the kitchen
d) Ensure there's a proper electrical outlet for the stove's ignition system

Tip: Take photos of the existing setup before you begin. They can be helpful references if you encounter any issues.

Troubleshooting Tip: If the gas line stub-out is in an inconvenient location, don't attempt to move it yourself. Consult a professional for any modifications to the existing gas line.

4. Install the Shut-off Valve

If there isn't already a shut-off valve near the stove location:

a) Turn off the main gas supply
b) Install a shut-off valve on the gas line stub-out
c) Use pipe joint compound on the threads
d) Tighten securely with a wrench

Tip: Position the shut-off valve so it will be easily accessible behind or beside the stove.

Troubleshooting Tip: If the existing stub-out doesn't allow for easy valve installation, you may need to add a short pipe extension first. Always use proper fittings and sealants.

5. Connect the Flexible Gas Connector

Assuming your local code allows flexible connectors:

a) Attach one end of the connector to the shut-off valve
b) Use pipe joint compound on the threads
c) Tighten securely, but be careful not to over-tighten

Tip: Always use a new flexible connector. Never reuse an old one, even if it looks fine.

Troubleshooting Tip: If the connector doesn't fit properly, don't force it or try to modify it. You may need a different size or type of connector.

6. Position the Stove

Now it's time to bring in your new appliance:

a) Carefully move the stove into position
b) Leave enough space to work behind it
c) Make sure it's level, adjusting the feet if necessary

Tip: Use a piece of cardboard under the stove to protect your floor while you work.

Troubleshooting Tip: If the stove doesn't fit or can't be leveled properly, double-check your measurements and floor levelness. You may need to make adjustments to the space.

7. Connect the Stove

Now for the crucial connection:

a) Locate the gas inlet on the stove
b) Connect the free end of the flexible connector to the stove's gas inlet
c) Use pipe joint compound on the threads
d) Tighten securely

Tip: Follow the stove manufacturer's instructions carefully. They may have specific requirements for connection.

Troubleshooting Tip: If the connection doesn't align properly, don't bend or force the flexible connector. You may need to adjust the stove's position or use a different length connector.

8. Test for Leaks

Before using the stove:

a) Turn on the gas supply at the shut-off valve
b) Apply a soapy water solution or use a gas leak detector on all connections
c) Look for bubbles or listen for the detector's alarm
d) If you find any leaks, turn off the gas, tighten connections, and retest

Tip: Be thorough in your leak check. Test every connection, no matter how secure you think it is.

Troubleshooting Tip: If you can't stop a leak by tightening, don't keep trying. Turn off the gas and disassemble the connection. Clean the threads, apply new sealant, and reconnect.

9. Connect the Electrical Supply

Most gas stoves need electricity for the ignition system:

a) Plug the stove into the nearby outlet
b) If hardwiring is required, follow the manufacturer's instructions carefully

Tip: Ensure the electrical cord isn't in contact with the back of the stove, which can get hot.

Troubleshooting Tip: If the plug doesn't fit the outlet or the cord doesn't reach, don't use an extension cord or adapter. You may need to have a new outlet installed by an electrician.

10. Test the Stove

Finally, it's time to test your installation:

a) Turn on each burner to ensure it lights properly
b) Check that the oven ignites correctly
c) Listen for any unusual sounds or smell for gas

Tip: Refer to your stove's manual for any model-specific testing procedures.

Troubleshooting Tip: If a burner won't light, check that its parts are correctly assembled. If the problem persists, consult the troubleshooting section of your manual or contact the manufacturer.

11. Final Adjustments and Safety Checks

Before considering the job complete:

a) Ensure the stove is still level
b) Check that all burner flames are blue and steady
c) Verify that the oven door seals properly
d) Make sure the stove's anti-tip bracket is securely installed

Tip: Take time to familiarize yourself with all the stove's features and safety mechanisms.

Troubleshooting Tip: If you notice yellow flames or any other issues, don't ignore them. These could indicate serious problems that need professional attention.

Remember, while installing a gas stove can be a rewarding DIY project, safety is paramount. If at any point you feel unsure or uncomfortable, don't hesitate to call in a professional. The peace of mind that comes with a safely installed appliance is well worth it.

Congratulations on installing your new gas stove! In our next section, we'll explore the process of connecting a gas fireplace, another popular and cozy addition to many homes. Are you ready to bring the warmth and ambiance of a gas fireplace into your living space? Let's move on to this exciting project!

Fireplace Connection

Imagine this: It's a chilly evening, and you're looking forward to curling up in front of your newly installed gas fireplace. The ambiance, the warmth, the convenience—it all sounds perfect. But before you can enjoy that cozy scene, you need to safely connect your fireplace to your gas line. This task might seem daunting, but with the right approach, you can do it confidently and safely.

When I first connected a gas fireplace, I was both excited and nervous. The prospect of potential gas leaks or improper installation was intimidating. However, I've learned that with careful attention to detail and following proper procedures, connecting a gas fireplace can be a rewarding DIY project. Let me walk you through the process, sharing insights I've gained to help you successfully and safely connect your gas fireplace.

1. Safety First

Before we begin, let's review crucial safety measures:

- Ensure proper ventilation in the work area
- Have a fire extinguisher nearby
- Know the location of your main gas shut-off valve
- Never smoke or have open flames in the area while working
- Wear safety glasses and work gloves

Tip: If you smell gas at any point during the installation, stop immediately, turn off the main gas supply, and call a professional.

Troubleshooting Tip: If you're unsure about your home's gas system or the fireplace's compatibility, consider having a professional inspection before starting the project.

2. Gather Your Tools and Materials

You'll need:

- Adjustable wrench
- Pipe wrench
- Gas leak detector or soapy water solution
- Flexible gas connector (if allowed by local codes)
- Black iron pipe and fittings (if rigid connection is required)
- Pipe joint compound (gas-rated)
- Teflon tape (gas-rated)
- Shut-off valve
- Sediment trap

Tip: Always use tools and materials specifically rated for gas line work.

Troubleshooting Tip: If your local code doesn't allow flexible connectors, you'll need to use black iron pipe. This requires more skill and tools, so consider professional help if you're not comfortable with pipe fitting.

3. Prepare the Area

Before bringing in the fireplace:

a) Ensure the fireplace location meets all clearance requirements specified by the manufacturer
b) Verify that proper venting is in place (for vented fireplaces)
c) Locate the gas line stub-out in the fireplace area
d) Clean the area thoroughly

Tip: Take photos of the existing setup before you begin. They can be helpful references if you encounter any issues.

Troubleshooting Tip: If the gas line stub-out is in an inconvenient location, don't attempt to move it yourself. Consult a professional for any modifications to the existing gas line.

4. Install the Shut-off Valve

A dedicated shut-off valve is crucial for safety:

a) Turn off the main gas supply
b) Install a shut-off valve on the gas line stub-out
c) Use pipe joint compound on the threads
d) Tighten securely with a wrench

Tip: Position the shut-off valve so it will be easily accessible, but not visible when the fireplace is in use.

Troubleshooting Tip: If the valve doesn't tighten properly or you notice any damage to the threads, replace the valve. Never use a faulty valve in a gas line.

5. Install the Sediment Trap

A sediment trap helps prevent debris from entering the fireplace:

a) Install a nipple extending downward from the shut-off valve
b) Attach a tee fitting to the nipple
c) Add a 3-inch nipple and cap to the bottom of the tee
d) The side outlet of the tee will connect to the fireplace

Tip: The sediment trap should be the last component before the gas enters the fireplace.

Troubleshooting Tip: If you're tight on space, you can use a shorter nipple for the trap, but never less than 3 inches.

6. Connect the Gas Line to the Fireplace

Depending on your local codes and fireplace model, you'll use either a flexible connector or rigid pipe:

For Flexible Connector:
a) Attach one end to the sediment trap tee
b) Route the connector to the fireplace's gas inlet
c) Attach the other end to the fireplace
d) Use pipe joint compound on all threads

For Rigid Pipe:
a) Measure and cut black iron pipe to fit from the sediment trap to the fireplace
b) Thread the pipe ends and use appropriate fittings
c) Use pipe joint compound on all threads
d) Tighten all connections securely

Tip: If using a flexible connector, ensure it's not kinked or stretched tight.

Troubleshooting Tip: If the connections don't align properly, don't force them. Recheck your measurements and adjust the pipe routing if necessary.

7. Position the Fireplace

Now it's time to move the fireplace into its final position:

a) Carefully slide the fireplace into place
b) Ensure all clearances are maintained as per manufacturer's specifications
c) Check that the gas connection is not strained or kinked

Tip: Use furniture sliders under the fireplace to make positioning easier and protect your floor.

Troubleshooting Tip: If the fireplace doesn't fit as expected, double-check all measurements. You may need to adjust the surrounding structure or choose a different fireplace model.

8. Test for Leaks

Before using the fireplace:

a) Turn on the gas supply at the shut-off valve
b) Apply a soapy water solution or use a gas leak detector on all connections
c) Look for bubbles or listen for the detector's alarm
d) If you find any leaks, turn off the gas, tighten connections, and retest

Tip: Be thorough in your leak check. Test every connection, no matter how secure you think it is.

Troubleshooting Tip: If you can't stop a leak by tightening, don't keep trying. Turn off the gas, disassemble the connection, clean the threads, apply new sealant, and reconnect.

9. Connect the Electrical Supply (if applicable)

Many gas fireplaces require electricity for ignition or fans:

a) Ensure power is off at the circuit breaker
b) Connect the fireplace to the electrical supply as per manufacturer's instructions
c) If hardwiring is required, consider hiring an electrician

Tip: Keep electrical connections away from hot surfaces of the fireplace.

Troubleshooting Tip: If the fireplace doesn't have power after connection, check the circuit breaker and verify all connections before calling an electrician.

10. Install Fireplace Media and Glass

Follow the manufacturer's instructions to:

a) Install lava rocks, glass beads, or ceramic logs
b) Ensure proper placement for optimal flame pattern
c) Install the glass front, if applicable

Tip: Take your time arranging the media. The right arrangement can significantly enhance the fireplace's appearance.

Troubleshooting Tip: If the flame pattern looks unusual after installation, check the media arrangement. Improper placement can affect gas flow and flame appearance.

11. Test the Fireplace

Finally, it's time to test your installation:

a) Turn on the gas and ignite the fireplace
b) Check for proper flame color and pattern
c) Verify that all safety features are working (e.g., oxygen depletion sensor)
d) Test any additional features like remote controls or thermostats

Tip: Let the fireplace run for at least 30 minutes during the first use, checking periodically for any issues.

Troubleshooting Tip: If you notice any unusual smells, sounds, or flame patterns, turn off the fireplace and consult the troubleshooting section of your manual or contact a professional.

12. Final Safety Checks

Before considering the job complete:

a) Ensure all clearances are still maintained after final positioning
b) Check that the area around the fireplace doesn't get excessively hot
c) Verify that carbon monoxide detectors are installed and working in the room
d) Review proper operation and safety procedures with all household members

Tip: Create a maintenance schedule for your fireplace, including annual professional inspections.

Troubleshooting Tip: If you have any lingering doubts about the installation or operation, don't hesitate to have a professional inspect your work. Safety should always be the top priority.

Remember, while connecting a gas fireplace can be a satisfying DIY project, it involves working with potentially dangerous gas and high temperatures. If at any point you feel unsure or uncomfortable, it's wise to call in a professional. The peace of mind and safety of a correctly installed fireplace are well worth it.

Congratulations on connecting your new gas fireplace! You're now ready to enjoy cozy evenings by the fire. In our next section, we'll explore the process of setting up an outdoor gas grill, perfect for those summer barbecues. Are you ready to take your culinary skills outdoors? Let's move on to this exciting addition to your home!

Outdoor Grill Setup

Picture this: It's a beautiful summer evening, and you're excited to fire up your new outdoor gas grill for a family barbecue. The aroma of grilled food, the convenience of instant heat—it all sounds perfect. But before you can become the neighborhood grill master, you need to safely set up and connect your outdoor grill to your gas line. This task might seem challenging, but with the right approach, you can do it confidently and safely.

When I first set up an outdoor gas grill, I was thrilled about the prospect of year-round grilling but also concerned about working with gas lines outdoors. Over time, I've learned that with careful planning and attention to safety, setting up an outdoor grill can be a rewarding DIY project. Let me guide you through the process, sharing insights I've gained to help you successfully and safely set up your outdoor gas grill.

1. Safety First

Before we begin, let's review crucial safety measures:

- Ensure the grill location is well-ventilated and away from structures
- Have a fire extinguisher nearby
- Know the location of your main gas shut-off valve
- Never smoke or have open flames in the area while working
- Wear safety glasses and work gloves

Tip: If you smell gas at any point during the installation, stop immediately, turn off the main gas supply, and call a professional.

Troubleshooting Tip: If you're unsure about your home's outdoor gas system, consider having a professional inspection before starting the project.

2. Gather Your Tools and Materials

You'll need:

- Adjustable wrench
- Pipe wrench
- Gas leak detector or soapy water solution
- Flexible gas connector (outdoor-rated)
- Black iron pipe and fittings (for the fixed portion of the line)
- Pipe joint compound (gas-rated)
- Teflon tape (gas-rated)
- Shut-off valve (outdoor-rated)
- Sediment trap
- Weather-resistant cover for connections

Tip: Always use tools and materials specifically rated for outdoor gas line work.

Troubleshooting Tip: If you're unsure about which materials are suitable for outdoor use, consult with a professional at your local hardware store or a licensed plumber.

3. Plan Your Grill Location

Choosing the right location is crucial:

a) Ensure the grill is at least 10 feet away from any structure
b) Choose a level surface that can support the grill's weight
c) Avoid placing the grill under eaves, overhanging branches, or enclosed spaces
d) Consider wind patterns to prevent smoke from blowing into your home

Tip: Think about your cooking workflow. Place the grill in a convenient location relative to your kitchen and outdoor dining area.

Troubleshooting Tip: If you can't find an ideal location that meets all criteria, consider building a dedicated grilling station that addresses safety concerns.

4. Run the Gas Line

Extend your home's gas line to the grill location:

a) Turn off the main gas supply
b) Dig a trench for the gas line (typically 18 inches deep, but check local codes)
c) Install black iron pipe from your home's existing gas line to the grill location
d) Use appropriate fittings and pipe joint compound at all connections
e) Rise out of the ground with a 90-degree elbow and a short vertical pipe

Tip: Call your local utility locating service before digging to avoid hitting other utility lines.

Troubleshooting Tip: If you encounter unexpected obstacles while digging, don't force it. You may need to adjust your route or consult a professional.

5. Install the Shut-off Valve

A dedicated outdoor shut-off valve is crucial for safety:

a) Install a shut-off valve at the end of the vertical pipe

b) Use pipe joint compound on the threads

c) Ensure the valve is easily accessible but protected from accidental damage

Tip: Consider installing the valve in a small, weather-resistant enclosure for protection and aesthetic purposes.

Troubleshooting Tip: If the valve feels stiff or doesn't turn smoothly, don't force it. Replace it with a new, high-quality outdoor-rated valve.

6. Install the Sediment Trap

A sediment trap helps prevent debris from entering the grill:

a) Install a tee fitting after the shut-off valve

b) Add a 3-inch nipple and cap pointing downward from the tee

c) The side outlet of the tee will connect to the grill

Tip: The sediment trap should be the last component before the flexible connector to the grill.

Troubleshooting Tip: If you notice a lot of debris when installing the sediment trap, consider having your entire gas line system inspected and cleaned.

7. Connect the Flexible Gas Line

Use an outdoor-rated flexible gas connector to allow for grill movement:

a) Attach one end of the connector to the sediment trap tee

b) Route the connector to the grill's gas inlet

c) Use pipe joint compound on all threads

d) Tighten all connections securely

Tip: Ensure the flexible connector is long enough to allow for grill movement but not so long that it could be damaged or tripped over.

Troubleshooting Tip: If the flexible connector kinks or doesn't reach comfortably, don't force it. You may need a different length or to adjust the grill's position.

8. Position the Grill

Now it's time to move the grill into its final position:

a) Carefully place the grill in the chosen location
b) Ensure it's level and stable
c) Check that the gas connection is not strained or kinked

Tip: Use a level to ensure the grill is perfectly horizontal for even cooking.

Troubleshooting Tip: If the grill isn't stable, consider installing it on a flat paving stone or building a dedicated grill pad.

9. Test for Leaks

Before using the grill:

a) Turn on the gas supply at the shut-off valve
b) Apply a soapy water solution or use a gas leak detector on all connections
c) Look for bubbles or listen for the detector's alarm
d) If you find any leaks, turn off the gas, tighten connections, and retest

Tip: Be extra thorough in your leak check for outdoor installations, as they're exposed to the elements.

Troubleshooting Tip: If you can't stop a leak by tightening, don't keep trying. Turn off the gas, disassemble the connection, clean the threads, apply new sealant, and reconnect.

10. Install Grill Components

Follow the manufacturer's instructions to:

a) Install burners, heat plates, and grates
b) Connect the ignition system if separate from the main unit
c) Attach side tables or other accessories

Tip: Take photos as you install components to help with future maintenance or cleaning.

Troubleshooting Tip: If any parts don't fit correctly, double-check the manual. Don't force anything, as this could lead to gas leaks or improper operation.

11. Test the Grill

Finally, it's time to test your installation:

a) Turn on the gas and ignite each burner
b) Check for proper flame color and pattern (should be mostly blue with slight yellow tips)
c) Verify that all burners light correctly, including side burners if present
d) Test any additional features like rotisserie or smoke box

Tip: Let the grill run for at least 15 minutes during the first use, checking periodically for any issues.

Troubleshooting Tip: If burners won't light or the flame looks unusual, check that the burners and gas ports are clean and unobstructed. If problems persist, consult your manual or a professional.

12. Final Safety Checks and Weatherproofing

Before considering the job complete:

a) Ensure all connections are protected from the weather (use covers if necessary)
b) Verify that the area around the grill remains clear of flammable materials
c) Test the shut-off valve to ensure it's easily operable
d) Consider installing a carbon monoxide detector in adjacent enclosed spaces

Tip: Create a maintenance schedule for your grill, including regular cleaning and annual inspections of the gas connections.

Troubleshooting Tip: If you notice any rust or corrosion on fittings during future use, replace them immediately. Outdoor installations require extra vigilance due to weather exposure.

Remember, while setting up an outdoor gas grill can be a great DIY project, it involves working with gas in an exposed environment. If at any point you feel unsure or uncomfortable, it's wise to call in a professional. The safety of your home and family is paramount.

Congratulations on setting up your outdoor gas grill! You're now ready to enjoy delicious barbecues all year round. In our next section, we'll discuss general safety considerations for all gas appliances. Are you ready to ensure your gas-powered home remains safe and efficient? Let's move on to this crucial topic!

Water Heater and Furnace Hookups

Imagine this: It's the dead of winter, and you're installing a new high-efficiency gas furnace and water heater. The promise of lower energy bills and endless hot water is exciting, but the thought of connecting these crucial appliances to your gas line might be a bit overwhelming. Don't worry! With the right approach and attention to detail, you can safely and effectively hook up your water heater and furnace.

When I first tackled these installations, I was nervous about the complexity and the potential risks. However, I've learned that with careful planning and adherence to safety protocols, connecting a water heater and furnace can be a rewarding DIY project. Let me guide you through this process, sharing insights I've gained to help you successfully and safely hook up these essential home appliances.

1. Safety First

Before we begin, let's review crucial safety measures:

- Ensure proper ventilation in the work area
- Have a fire extinguisher nearby
- Know the location of your main gas shut-off valve
- Never smoke or have open flames in the area while working
- Wear safety glasses and work gloves
- Consider having a carbon monoxide detector in the area

Tip: If you smell gas at any point during the installation, stop immediately, turn off the main gas supply, and call a professional.

Troubleshooting Tip: If you're unsure about your home's gas system or the compatibility of your new appliances, consider having a professional inspection before starting the project.

2. Gather Your Tools and Materials

You'll need:

- Adjustable wrench
- Pipe wrench
- Gas leak detector or soapy water solution
- Flexible gas connectors (if allowed by local codes)
- Black iron pipe and fittings (if rigid connection is required)
- Pipe joint compound (gas-rated)
- Teflon tape (gas-rated)
- Shut-off valves
- Sediment traps
- Appropriate venting materials (for the furnace and water heater)

Tip: Always use tools and materials specifically rated for gas line work.

Troubleshooting Tip: If your local code doesn't allow flexible connectors, you'll need to use black iron pipe. This requires more skill and tools, so consider professional help if you're not comfortable with pipe fitting.

3. Prepare the Installation Areas

Before bringing in the new appliances:

a) Ensure proper clearances as specified by the manufacturers
b) Verify that proper venting is in place or can be installed
c) Locate the gas line stub-outs in each area
d) Clean the areas thoroughly

Tip: Take photos of the existing setups before you begin. They can be helpful references if you encounter any issues.

Troubleshooting Tip: If the gas line stub-outs are in inconvenient locations, don't attempt to move them yourself. Consult a professional for any modifications to the existing gas lines.

4. Install Shut-off Valves

For both the water heater and furnace:

a) Turn off the main gas supply
b) Install a shut-off valve on each gas line stub-out
c) Use pipe joint compound on the threads
d) Tighten securely with a wrench

Tip: Position the shut-off valves so they will be easily accessible, but not in a way that interferes with the appliance installation.

Troubleshooting Tip: If a valve doesn't tighten properly or you notice any damage to the threads, replace the valve. Never use a faulty valve in a gas line.

5. Install Sediment Traps

For both appliances:

a) Install a nipple extending downward from each shut-off valve
b) Attach a tee fitting to each nipple
c) Add a 3-inch nipple and cap to the bottom of each tee
d) The side outlet of each tee will connect to its respective appliance

Tip: Sediment traps should be the last components before the gas enters each appliance.

Troubleshooting Tip: If you're tight on space, you can use shorter nipples for the traps, but never less than 3 inches.

6. Connect the Gas Lines to the Appliances

Depending on your local codes and appliance models, you'll use either flexible connectors or rigid pipe:

For Flexible Connectors:
a) Attach one end to the sediment trap tee
b) Route the connector to the appliance's gas inlet
c) Attach the other end to the appliance
d) Use pipe joint compound on all threads

For Rigid Pipe:
a) Measure and cut black iron pipe to fit from the sediment trap to the appliance
b) Thread the pipe ends and use appropriate fittings
c) Use pipe joint compound on all threads
d) Tighten all connections securely

Tip: If using flexible connectors, ensure they're not kinked or stretched tight.

Troubleshooting Tip: If the connections don't align properly, don't force them. Recheck your measurements and adjust the pipe routing if necessary.

7. Install Venting

Proper venting is crucial for both water heaters and furnaces:

a) Follow manufacturer instructions for venting requirements
b) Ensure proper slope on horizontal runs (typically 1/4 inch per foot)
c) Use appropriate materials (e.g., double-wall for high-efficiency appliances)
d) Seal all joints and connections

Tip: Consider hiring a professional for this step if you're not experienced with venting systems. Improper venting can lead to dangerous carbon monoxide buildup.

Troubleshooting Tip: If you encounter obstacles while routing the vent, don't compromise on proper installation. Rethink the appliance location or consult a professional for solutions.

8. Connect Water Lines (for Water Heater)

For the water heater:

a) Install flexible water connectors to the hot and cold water ports
b) Connect these to your home's water supply lines
c) Use Teflon tape on threaded connections

Tip: Consider installing water shut-off valves if not already present.

Troubleshooting Tip: If you notice any leaks after turning on the water, tighten connections carefully. If leaks persist, you may need to replace the connectors or check for damaged threads.

9. Electrical Connections

Both appliances will likely require electrical connections:

a) Ensure power is off at the circuit breaker
b) Connect the appliances to their respective electrical supplies as per manufacturer's instructions
c) If hardwiring is required, consider hiring an electrician

Tip: Keep electrical connections away from hot surfaces and water sources.

Troubleshooting Tip: If an appliance doesn't have power after connection, check the circuit breaker and verify all connections before calling an electrician.

10. Test for Gas Leaks

Before using the appliances:

a) Turn on the gas supply at the shut-off valves
b) Apply a soapy water solution or use a gas leak detector on all gas connections
c) Look for bubbles or listen for the detector's alarm
d) If you find any leaks, turn off the gas, tighten connections, and retest

Tip: Be thorough in your leak check. Test every connection, no matter how secure you think it is.

Troubleshooting Tip: If you can't stop a leak by tightening, don't keep trying. Turn off the gas, disassemble the connection, clean the threads, apply new sealant, and reconnect.

11. Test the Appliances

Finally, it's time to test your installations:

a) Follow the manufacturer's instructions to light the pilot lights or initiate electronic ignition
b) Run both appliances through a complete cycle
c) Check for proper flame color and pattern (should be mostly blue)
d) Listen for any unusual noises
e) For the water heater, verify hot water production
f) For the furnace, check for proper airflow and heating

Tip: Let each appliance run for at least 30 minutes during the first use, checking periodically for any issues.

Troubleshooting Tip: If you notice any unusual smells, sounds, or operation, turn off the appliance and consult the troubleshooting section of your manual or contact a professional.

12. Final Safety Checks

Before considering the job complete:

a) Ensure all clearances are still maintained after final positioning
b) Verify that carbon monoxide detectors are installed and working in nearby areas
c) Review proper operation and safety procedures for both appliances
d) Schedule a professional inspection if required by local codes

Tip: Create a maintenance schedule for both appliances, including annual professional inspections.

Troubleshooting Tip: If you have any lingering doubts about the installations or operation, don't hesitate to have a professional inspect your work. Safety should always be the top priority.

Remember, while hooking up a water heater and furnace can be a satisfying DIY project, it involves working with gas, electricity, and potentially dangerous exhaust gases. If at any point you feel unsure or uncomfortable, it's wise to call in a professional. The peace of mind and safety of correctly installed appliances are well worth it.

Congratulations on hooking up your new water heater and furnace! You've taken a significant step in improving your home's efficiency and comfort. In our next section, we'll discuss troubleshooting common gas line issues. Are you ready to become a gas line problem-solving expert? Let's move on to this essential skill set!

Chapter 6
Troubleshooting and Repairs
Detecting Gas Leaks

Imagine this scenario: You're relaxing at home when you notice a faint, unusual odor. Your mind races – could it be a gas leak? The ability to detect and respond to gas leaks quickly is a crucial skill for any homeowner. It's not just about maintaining your gas system; it's about ensuring the safety of your home and loved ones.

When I first encountered a suspected gas leak, I felt a mix of anxiety and uncertainty. Over time, I've learned that with the right knowledge and tools, detecting gas leaks can be approached calmly and systematically. Let me guide you through the process of detecting gas leaks, sharing insights I've gained to help you handle this potentially dangerous situation with confidence.

1. Understanding Gas Leak Indicators

Before we dive into detection methods, let's review the signs of a potential gas leak:

- Smell: Natural gas and propane have an added "rotten egg" odor for easy detection.
- Sound: Hissing or whistling noises near gas lines or appliances.
- Sight: Dead or discolored vegetation around gas lines, or bubbles in standing water.
- Physical symptoms: Dizziness, nausea, or headaches when in your home.

Tip: Familiarize all household members with these signs and what to do if they suspect a leak.

Troubleshooting Tip: If you or family members experience sudden, unexplained physical symptoms at home, don't ignore them. Evacuate and call emergency services – it could be carbon monoxide or a gas leak.

2. Using Your Nose

Your sense of smell is often the first line of defense:

a) Regularly "sniff test" areas around gas appliances and visible gas lines.
b) Pay attention to any unusual odors, especially the distinct "rotten egg" smell.
c) Remember, not all gas leaks produce a strong odor immediately.

Tip: If you have any doubts about your ability to smell gas (due to age, medical conditions, etc.), consider installing gas detectors in your home.

Troubleshooting Tip: If you become "nose blind" to the gas smell after prolonged exposure, leave the area for fresh air and return. If you still can't smell it but suspect a leak, use other detection methods.

3. Visual Inspection

Regularly inspect your gas lines and appliances:

a) Look for physical damage to pipes or connections.
b) Check for rust or corrosion on metal parts.
c) Inspect flexible gas lines for cracks or signs of wear.
d) Watch for unusual flame color in gas appliances (should be blue, not yellow or orange).

Tip: Create a schedule for visual inspections, perhaps coinciding with changing your smoke detector batteries.

Troubleshooting Tip: If you notice any physical damage, no matter how small, don't ignore it. Even a tiny crack can lead to a dangerous leak over time.

4. Using Soapy Water Solution

This simple yet effective method can help locate the source of a small leak:

a) Mix a solution of dish soap and water in a spray bottle.
b) Spray the solution on gas connections and along gas lines.
c) Watch for bubbles forming, which indicate a leak.

Tip: Use a flashlight to help spot small bubbles more easily.

Troubleshooting Tip: If you can't see bubbles but still suspect a leak, try wiping the area with a paper towel – it might pick up soap residue from tiny bubbles you can't see.

5. Electronic Gas Detectors

For more precise detection, use an electronic gas detector:

a) Choose a detector that can sense both natural gas and propane.
b) Turn on the detector in a gas-free area to establish a baseline.
c) Slowly move the sensor along gas lines and around appliances.
d) Pay attention to any increase in the detector's readings.

Tip: Invest in a quality detector with both visual and audible alarms.

Troubleshooting Tip: If your detector gives inconsistent readings, check the batteries and recalibrate according to the manufacturer's instructions. If problems persist, replace the unit.

6. Professional Leak Detection Services

For thorough inspections or if you suspect a hard-to-find leak:

a) Contact your gas company or a licensed plumber for a professional inspection.
b) They may use advanced tools like ultrasonic detectors or gas imaging cameras.
c) Professional services can often detect leaks in hard-to-reach areas or underground pipes.

Tip: Consider scheduling professional inspections annually, especially if your gas system is older.

Troubleshooting Tip: If a professional can't find a leak but you still have concerns, don't hesitate to get a second opinion. Trust your instincts when it comes to safety.

7. Testing Gas Appliances

Regularly check your gas appliances for proper operation:

a) Listen for unusual noises when appliances are running.
b) Check for soot or scorching around burners or vents.
c) Pay attention to pilot lights that frequently go out or are difficult to light.
d) Watch for excessive condensation on windows or walls near appliances.

Tip: Keep appliance manuals handy and follow recommended maintenance schedules.

Troubleshooting Tip: If an appliance is giving you persistent trouble, it's often more cost-effective (and safer) to replace it rather than continuing to repair it.

8. Using Carbon Monoxide Detectors

While not specifically for gas leak detection, CO detectors are crucial for safety:

a) Install CO detectors on every level of your home and near sleeping areas.
b) Test detectors monthly and replace batteries as needed.
c) Replace the entire unit according to the manufacturer's recommendations (usually every 5-7 years).

Tip: Choose CO detectors with digital displays to monitor CO levels continuously.

Troubleshooting Tip: If your CO detector alarms, treat it as an emergency. Evacuate immediately and call emergency services. Don't assume it's a false alarm.

9. Educate Family Members

Ensure everyone in your household knows how to respond to a suspected gas leak:

a) Teach them the signs of a gas leak.
b) Establish an evacuation plan.
c) Show them how to turn off the main gas supply.
d) Emphasize the importance of not using any electrical devices or open flames if a leak is suspected.

Tip: Conduct periodic "gas leak drills" to keep everyone prepared.

Troubleshooting Tip: If family members are reluctant to participate in safety planning, share real-life stories about the dangers of gas leaks to emphasize the importance of preparedness.

10. Regular Maintenance

Preventing leaks is as important as detecting them:

a) Schedule annual inspections of your gas system and appliances.
b) Replace flexible gas lines every 10 years or sooner if they show signs of wear.
c) Keep outdoor gas lines clear of debris and protect them from physical damage.
d) Ensure proper ventilation for all gas appliances.

Tip: Keep a log of all maintenance and inspections for future reference.

Troubleshooting Tip: If you notice an increase in your gas bill without a corresponding increase in usage, it could indicate a slow leak. Don't ignore unexpected changes in your utility bills.

Remember, when it comes to gas leaks, it's always better to be safe than sorry. If you ever have doubts about your ability to safely detect or handle a potential gas leak, don't hesitate to call professionals. Your safety and that of your family is paramount.

In our next section, we'll discuss how to safely respond to a confirmed gas leak and the steps for basic repairs. Are you ready to learn how to handle gas emergencies with confidence? Let's move on to this crucial skill set!

Fixing Common Gas Line Issues

Imagine this: You've just detected a minor gas leak or noticed an issue with your gas line. Your heart might be racing, but don't panic! Many common gas line issues can be safely addressed with the right knowledge and precautions. However, it's crucial to understand your limits and know when to call in a professional.

When I first encountered gas line issues, I was tempted to tackle everything myself. Over time, I've learned which problems I can handle safely and which require expert intervention. Let me guide you through fixing common gas line issues, sharing insights I've gained to help you address these problems confidently and safely.

1. Safety First

Before attempting any repairs:

- Ensure the area is well-ventilated
- Have a fire extinguisher nearby
- Know where your main gas shut-off valve is located
- Never smoke or use any flame or spark-producing devices near gas lines
- Wear protective gear: safety glasses, gloves, and a dust mask

Tip: If you smell a strong gas odor or hear a loud hissing, evacuate immediately and call your gas company or emergency services.

Troubleshooting Tip: If you're ever in doubt about your ability to safely handle a gas line issue, don't hesitate to call a professional. Safety should always be your top priority.

Tip: If you find extensive corrosion, consider replacing a larger section of the gas line to prevent future issues.

2. Tightening Loose Connections

Loose connections are a common source of small leaks:

a) Turn off the gas supply to the affected area
b) Use two wrenches - one to hold the fitting steady and one to tighten
c) Apply pipe joint compound to the threads before tightening
d) Tighten until snug, but be careful not to over-tighten

Tip: Always use the correct size wrench to avoid damaging the fittings.

Troubleshooting Tip: If tightening doesn't stop the leak, the fitting may be damaged and need replacement. Don't keep tightening an already tight connection.

3. Replacing Faulty Shut-off Valves

A malfunctioning shut-off valve can be dangerous:

a) Turn off the gas supply upstream from the valve
b) Disconnect the valve using two wrenches to avoid twisting the pipe
c) Clean the pipe threads and apply new pipe joint compound
d) Install the new valve, ensuring it's oriented correctly for gas flow
e) Tighten securely and test for leaks

Tip: Choose a high-quality replacement valve rated for gas use.

Troubleshooting Tip: If the new valve doesn't align properly, you may need to adjust the pipe configuration. Don't force the valve into position.

4. Repairing Damaged Flexible Gas Lines

Flexible gas lines should be replaced if damaged:

a) Turn off the gas supply
b) Disconnect the flexible line at both ends
c) Measure and cut a new flexible line to the correct length
d) Install the new line, using new fittings if necessary
e) Tighten all connections and test for leaks

Tip: Never reuse old fittings with a new flexible line.

Troubleshooting Tip: If the new line doesn't fit properly, don't try to stretch or bend it forcefully. You may need a different length or configuration.

5. Addressing Corrosion

Corroded pipes or fittings need immediate attention:

a) Turn off the gas supply
b) Cut out the corroded section of pipe
c) Clean the remaining pipe ends thoroughly
d) Install a new section of pipe using appropriate fittings
e) Apply pipe joint compound and tighten all connections

Troubleshooting Tip: If corrosion is widespread, it may indicate a larger problem with your gas system. Consider a professional inspection of the entire system.

6. Fixing Leaky Pipe Threads

Leaky threads can often be fixed without replacing the entire fitting:

a) Turn off the gas supply
b) Disconnect the leaking joint
c) Clean the threads thoroughly with a wire brush
d) Apply new pipe joint compound or PTFE tape
e) Reconnect and tighten the joint

Tip: Use yellow PTFE tape specifically rated for gas lines, not the white tape used for water pipes.

Troubleshooting Tip: If the leak persists after resealing, the threads may be damaged. In this case, the fitting or pipe section may need replacement.

7. Addressing Gas Pressure Issues

Inconsistent gas pressure can cause appliance problems:

a) Check that the main gas valve is fully open
b) Inspect the gas meter to ensure it's functioning properly
c) Look for kinked or crushed gas lines that could restrict flow
d) Clean or replace dirty gas filters on appliances

Tip: If pressure problems persist, contact your gas company to check the supply pressure to your home.

Troubleshooting Tip: If only one appliance is affected, the issue may be with the appliance's gas valve or burner orifices. Consult the appliance manual or a professional for specific troubleshooting steps.

8. Repairing Small Pipe Punctures

For very small holes in accessible pipes:

a) Turn off the gas supply
b) Clean the area around the hole thoroughly
c) Apply epoxy putty specifically designed for gas line repair
d) Allow the epoxy to cure fully before turning the gas back on

Tip: This is a temporary fix. Plan to replace the damaged section of pipe as soon as possible.

Troubleshooting Tip: If the hole is larger than a pinprick or the pipe is under high pressure, do not attempt this repair. Call a professional immediately.

9. Addressing Gas Odors with No Apparent Leak

If you smell gas but can't find a leak:

a) Check pilot lights on all gas appliances
b) Inspect gas connections on recently moved appliances
c) Look for unused gas outlets that may not be properly capped
d) Consider that the odor might be from sewer gas, not natural gas

Tip: Some household products can smell similar to gas. Rule out other sources before assuming it's a gas leak.

Troubleshooting Tip: If you can't identify the source of the odor, don't take chances. Call your gas company for a professional inspection.

10. Upgrading Old or Outdated Gas Lines

If your gas lines are old or not up to current codes:

a) Consult with a licensed plumber or gas fitter
b) Obtain necessary permits for the work
c) Consider replacing old pipes with modern materials like CSST
d) Ensure all new installations meet current safety standards and local codes

Tip: This is often best left to professionals, especially if it involves significant changes to your gas system.

Troubleshooting Tip: If you decide to upgrade your system yourself, be prepared for unexpected issues, especially in older homes. Have a contingency plan and budget for professional help if needed.

Remember, while many common gas line issues can be addressed by a knowledgeable homeowner, gas systems are inherently dangerous. Never hesitate to call a professional if you're unsure about any repair. It's always better to err on the side of caution when dealing with gas lines.

In our next section, we'll discuss when it's absolutely necessary to call in a professional for gas line issues. Are you ready to learn about the limits of DIY gas line work and when expert help is crucial? Let's move on to this important topic!

When to Call a Professional

Picture this: You're faced with a gas line issue, and you're wondering whether to tackle it yourself or call in an expert. This decision can be crucial for your safety and the integrity of your home's gas system. While it's empowering to handle some gas line tasks yourself, it's equally important to recognize when a situation calls for professional expertise.

When I first started working on gas lines, I was eager to handle everything myself. However, I quickly learned that there are times when calling a professional isn't just the smart choice – it's the only safe choice. Let me guide you through understanding when to call a professional for gas line issues, sharing insights I've gained to help you make this critical decision confidently.

1. Strong Gas Odors or Hissing Sounds

If you encounter:
- A strong, persistent smell of gas
- Loud hissing sounds from gas lines or appliances

Action:
- Evacuate the premises immediately
- Call your gas company or emergency services from outside the building

Tip: Don't try to locate or fix the source of a major leak yourself. The risk of explosion or fire is too high.

Troubleshooting Tip: If you're unsure whether the smell is actually gas, err on the side of caution and call professionals. It's better to be safe than sorry.

2. After Natural Disasters or Home Damage

Situations include:
- Earthquakes, floods, or severe storms
- Any event that may have shifted your home's foundation
- Physical damage to your home that could have affected gas lines

Action:
- Have a professional inspect your entire gas system before using any gas appliances

Tip: Even if you don't see obvious damage, hidden issues could exist.

Troubleshooting Tip: If you smell gas after a disaster, treat it as an emergency and follow evacuation procedures immediately.

3. Complex Installation or Modification Projects

Examples include:
- Installing new gas lines or extending existing ones
- Relocating major gas appliances (e.g., moving a water heater)
- Converting from propane to natural gas or vice versa

Action:
- Consult with a licensed plumber or gas fitter
- Obtain necessary permits and inspections

Tip: These projects often require specialized tools and knowledge of local codes.

Troubleshooting Tip: If you've started a project and realize it's more complex than anticipated, don't hesitate to call a professional mid-project. They can often work with what you've done and complete the job safely.

4. Persistent or Recurring Issues

If you experience:
- Gas leaks that keep returning after DIY repairs
- Appliances that frequently malfunction or have inconsistent gas flow

Action:
- Call a professional to diagnose and address the root cause

Tip: Recurring issues often indicate a more serious underlying problem.

Troubleshooting Tip: Keep a log of issues and repairs. This information can be valuable for the professional in diagnosing the problem.

5. Outdated or Corroded Gas Systems

Signs include:
- Visible rust or corrosion on gas pipes
- Gas lines made of materials no longer considered safe (e.g., lead pipes)
- A system that hasn't been professionally inspected in over 10 years

Action:
- Schedule a comprehensive inspection and potential system upgrade

Tip: Updating an old system can improve safety and efficiency.

Troubleshooting Tip: If you're unsure about the age or condition of your gas system, a professional can provide an assessment and recommend necessary updates.

6. When Working with High-Pressure Gas Lines

Situations involving:
- Any lines with pressure higher than standard residential levels
- Commercial or industrial gas systems

Action:
- Always use certified professionals for these high-risk systems

Tip: High-pressure systems require specialized knowledge and equipment.

Troubleshooting Tip: If you're unsure about the pressure in your system, consult your gas company or a professional before attempting any work.

7. Legal or Insurance Requirements

Circumstances include:
- Work that requires permits or inspections
- Repairs or installations that may affect your home insurance
- Any gas work in rental properties or commercial buildings

Action:
- Hire a licensed professional to ensure compliance with all regulations

Tip: Improper DIY work can lead to legal issues or invalidate insurance policies.

Troubleshooting Tip: If you're unsure about the legal requirements for gas work in your area, contact your local building department for guidance.

8. When You Lack Proper Tools or Equipment

If you:
- Don't have specialized tools required for gas line work
- Lack proper safety equipment or gas detection devices

Action:
- Call a professional who has the necessary tools and equipment

Tip: Attempting gas work without proper tools can be dangerous and often leads to incomplete or faulty repairs.

Troubleshooting Tip: If you find yourself improvising or using makeshift tools, stop and call a professional. The risk isn't worth the potential savings.

9. Unexplained Increases in Gas Bills

If you notice:
- A sudden or gradual increase in gas consumption without apparent cause
- Bills that seem unusually high compared to neighbors or previous years

Action:
- Have a professional inspect for leaks and check the efficiency of your gas appliances

Tip: Increased bills can indicate leaks too small for you to detect but significant enough to impact consumption.

Troubleshooting Tip: Before calling a professional, rule out other factors like rate increases or changes in your usage patterns. Your gas company can often provide historical usage data to help.

10. When You Feel Uncomfortable or Uncertain

If you:
- Feel overwhelmed by the complexity of the issue
- Are unsure about any aspect of gas line repair or maintenance
- Have any doubts about your ability to complete the work safely

Action:
- Trust your instincts and call a professional

Tip: There's no shame in prioritizing safety and seeking expert help.

Troubleshooting Tip: If you start a project and begin to feel uncomfortable, stop immediately. It's better to call a professional midway than to continue with uncertainty.

11. Carbon Monoxide Concerns

If you experience:
- Carbon monoxide detector alarms
- Symptoms like headaches, dizziness, or nausea when gas appliances are running

Action:
- Evacuate immediately and call emergency services
- Have a professional inspect all gas appliances and venting systems

Tip: Carbon monoxide is odorless and extremely dangerous. Never ignore these signs.

Troubleshooting Tip: Even if symptoms subside when you leave the house, don't return until a professional has checked your home. CO poisoning can be cumulative and deadly.

Remember, working with gas lines carries inherent risks. While it's admirable to want to handle home repairs yourself, gas line work is an area where professional expertise can be crucial for your safety and peace of mind. When in doubt, always lean towards calling a professional. The cost of professional service is a small price to pay for the safety of your home and family.

In our next section, we'll discuss maintenance practices to keep your gas system in top condition and potentially avoid many of these issues. Are you ready to learn how to be proactive in maintaining your gas lines? Let's move on to this important aspect of gas line care!

Chapter 7
Maintenance and Safety Practices
Regular Inspection Routines

Imagine this: You're sipping your morning coffee, feeling secure in your home. But when was the last time you gave your gas system a thorough check-up? Regular inspections are like health check-ups for your home's gas lines – they catch small issues before they become big problems, ensuring your peace of mind and your family's safety.

When I first became a homeowner, I underestimated the importance of regular gas line inspections. Over time, I've learned that a proactive approach to maintenance can prevent many headaches and potential dangers. Let me guide you through setting up and performing regular inspection routines, sharing insights I've gained to help you keep your gas system in top shape.

1. Establishing an Inspection Schedule

Create a routine that works for you:

- Monthly: Quick visual checks and leak tests
- Quarterly: More thorough inspections of accessible components
- Annually: Comprehensive inspection, potentially with professional help

Tip: Set reminders on your phone or calendar to stay on track with your inspection schedule.

Troubleshooting Tip: If you find it hard to stick to a regular schedule, try tying your inspections to other routine events, like changing smoke detector batteries or the start of each season.

2. Monthly Quick Checks

Perform these simple checks each month:

a) Visual inspection of visible gas lines for any signs of damage or corrosion
b) Check for any unusual odors around gas appliances and lines
c) Listen for any hissing sounds near gas connections
d) Ensure gas appliances are operating normally (e.g., burner flames are blue)

Tip: Make a checklist and keep it handy to ensure you don't miss any steps.

Troubleshooting Tip: If you notice anything unusual during these quick checks, don't wait for your next scheduled inspection. Investigate further or call a professional immediately.

3. Quarterly In-Depth Inspections

Every three months, conduct a more thorough examination:

a) Inspect all accessible gas lines, including those in basements, attics, and crawl spaces
b) Check all gas connections and fittings for signs of wear or corrosion
c) Test all shut-off valves to ensure they operate smoothly
d) Examine the areas around gas lines for any signs of pest activity or water damage
e) Check outdoor gas lines for exposure to elements or physical damage

Tip: Use a flashlight to better see in dim areas, and don't forget to check behind and under appliances.

Troubleshooting Tip: If you find a valve that's stiff or difficult to turn, don't force it. Instead, call a professional to lubricate or replace it.

4. Annual Comprehensive Inspection

Once a year, perform a complete system check:

a) Conduct a thorough leak test on all accessible gas lines and connections
b) Inspect the gas meter for any signs of damage or tampering
c) Check the pressure regulator (if accessible) for proper operation
d) Examine all gas appliances, including checking for proper ventilation
e) Review your gas usage history for any unexplained increases
f) Consider having a professional inspection for a more thorough evaluation

Tip: Schedule your annual inspection for the same time each year, perhaps before the heating season begins.

Troubleshooting Tip: If you're not comfortable performing any part of the annual inspection yourself, it's worth investing in a professional inspection for peace of mind.

5. Leak Testing Techniques

Incorporate these leak detection methods into your routine:

a) Soap Solution Test:
- Mix dish soap with water in a spray bottle
- Spray the solution on gas connections and along gas lines
- Watch for bubbles, which indicate a leak

b) Electronic Gas Detector:
- Use a handheld gas detector to check for leaks
- Move the sensor slowly along gas lines and around connections
- Pay attention to any increase in the detector's readings

Tip: Always start with the main gas connection and work your way through the system systematically.

Troubleshooting Tip: If you get inconsistent readings with an electronic detector, check the batteries and recalibrate the device. If problems persist, consider replacing the unit.

6. Inspecting Outdoor Components

Don't forget about external gas lines and equipment:

a) Check for physical damage from lawn equipment, vehicles, or falling branches
b) Look for signs of corrosion, especially where pipes enter the ground
c) Ensure proper protection is in place for above-ground pipes
d) Clear away any vegetation growing too close to gas lines

Tip: After severe weather, perform an additional check on outdoor gas components.

Troubleshooting Tip: If you notice any shifting or settling of the ground around gas lines, have a professional inspect for potential stress on the pipes.

7. Documenting Your Inspections

Keep detailed records of your inspections:

a) Create a log book or digital document for each inspection
b) Note the date, areas inspected, and any findings
c) Take photos of any concerning areas for future reference
d) Record any repairs or professional services performed

Tip: Use a consistent format for your records to make it easy to track changes over time.

Troubleshooting Tip: If you notice a trend of recurring issues in your logs, it may indicate a larger problem that needs professional attention.

8. Educating Household Members

Involve your family in the inspection process:

a) Teach them the signs of a potential gas leak
b) Show them how to perform basic visual checks
c) Ensure everyone knows where the main gas shut-off valve is located
d) Establish an emergency plan in case of a gas leak

Tip: Make it a family activity to promote safety awareness and shared responsibility.

Troubleshooting Tip: If family members are reluctant to participate, share real-life stories about the importance of gas safety to emphasize why their involvement matters.

9. Seasonal Considerations

Adjust your inspection routine based on the season:

a) Spring: Check for any damage from winter freezes or thaws
b) Summer: Inspect outdoor grills and lines before the barbecue season
c) Fall: Prepare heating systems for winter use
d) Winter: Monitor for ice buildup on outdoor components

Tip: Use the change of seasons as a reminder to perform your more thorough inspections.

Troubleshooting Tip: If you live in an area with extreme weather, consider more frequent checks during challenging seasons.

10. Professional Assessments

Complement your DIY inspections with professional help:

a) Schedule a professional inspection at least every 2-3 years
b) Have a professional check any areas you can't safely access
c) Consult experts for any persistent issues or concerns

Tip: Build a relationship with a trusted gas professional for consistent service and advice.

Troubleshooting Tip: If a professional finds issues you've missed in your inspections, ask them to show you what to look for to improve your own inspection skills.

Remember, regular inspections are your first line of defense against gas-related hazards. They not only ensure the safety of your home and family but can also save you money by catching small issues before they become costly repairs. While it might seem like a lot of work, the peace of mind that comes from knowing your gas system is in good condition is invaluable.

In our next section, we'll discuss cleaning and upkeep practices to maintain your gas system's efficiency and longevity. Are you ready to learn how to keep your gas lines and appliances in top condition? Let's move on to these essential maintenance tasks!

Cleaning and Upkeep

Imagine this: Your gas appliances are running smoothly, your energy bills are reasonable, and you're feeling confident about your home's safety. This scenario is achievable with proper cleaning and upkeep of your gas system. Just like any other part of your home, gas lines and appliances need regular care to function at their best.

When I first started maintaining my home's gas system, I was surprised by how much difference regular cleaning and upkeep could make. Over time, I've learned that these simple tasks not only extend the life of your gas system but also enhance its safety and efficiency. Let me guide you through the essential cleaning and upkeep practices, sharing insights I've gained to help you maintain your gas system like a pro.

1. Cleaning Gas Appliance Burners

Regular cleaning of burners ensures efficient operation:

a) Turn off the gas supply and allow the appliance to cool completely
b) Remove burner grates and caps
c) Use a soft brush or vacuum to remove loose debris
d) For stubborn grime, use a mixture of warm water and mild detergent
e) Dry thoroughly before reassembling

Tip: Always refer to your appliance's manual for specific cleaning instructions.

Troubleshooting Tip: If burners are producing yellow flames even after cleaning, the gas ports might be clogged. Use a thin wire to gently clear any blockages.

2. Maintaining Gas Stoves and Ovens

Keep your cooking appliances in top shape:

a) Clean spills immediately to prevent buildup
b) Regularly check and clean the oven's gas igniter
c) Inspect door seals for damage or wear
d) Clean or replace the oven's grease filter if applicable

Tip: Use a toothbrush to clean hard-to-reach areas around burners and controls.

Troubleshooting Tip: If your oven temperature seems inaccurate, check the calibration. Many ovens have a simple calibration process outlined in the manual.

3. Caring for Gas Furnaces

Proper furnace maintenance is crucial for safety and efficiency:

a) Replace or clean the air filter every 1-3 months
b) Vacuum around the furnace to remove dust and debris
c) Check and clean the blower assembly annually
d) Inspect the heat exchanger for cracks or corrosion
e) Keep the area around the furnace clear of clutter

Tip: Mark your calendar to check the furnace filter monthly during heavy use seasons.

Troubleshooting Tip: If your furnace is short-cycling (turning on and off frequently), a dirty filter might be the culprit. Replace it and see if the problem resolves.

4. Maintaining Gas Water Heaters

Extend the life of your water heater with these steps:

a) Flush the tank annually to remove sediment buildup
b) Check the anode rod every 2-3 years and replace if necessary
c) Inspect the pressure relief valve annually
d) Keep the area around the water heater clear and dry

Tip: When flushing the tank, let the water run until it's clear to ensure all sediment is removed.

Troubleshooting Tip: If you hear popping or rumbling sounds from your water heater, it's likely due to sediment buildup. Flush the tank as soon as possible to prevent damage.

5. Cleaning Gas Fireplaces

Keep your fireplace safe and attractive:

a) Turn off the gas and allow the unit to cool completely
b) Vacuum loose debris from logs and burners
c) Clean the glass with a non-ammonia glass cleaner
d) Inspect and clean the blower if your unit has one
e) Check for any soot buildup, which could indicate incomplete combustion

Tip: Take a photo of your gas logs before removing them for cleaning to ensure proper reassembly.

Troubleshooting Tip: If you notice excessive soot buildup, have a professional check the gas mixture and ventilation. This could indicate a potentially dangerous situation.

6. Maintaining Outdoor Gas Lines

Protect exposed gas lines from the elements:

a) Clear vegetation and debris from around outdoor gas lines
b) Check for signs of corrosion, especially where pipes enter the ground
c) Ensure protective coatings are intact on above-ground pipes
d) In cold climates, prevent ice buildup on outdoor gas components

Tip: Consider installing protective covers for outdoor gas meters and regulators.

Troubleshooting Tip: If you notice peeling paint or rust on outdoor gas pipes, clean the area gently and apply a gas-pipe-specific protective coating.

7. Cleaning Gas Line Exteriors

While you shouldn't attempt to clean the inside of gas lines, keeping the exteriors clean is important:

a) Wipe down visible gas pipes with a dry cloth to remove dust
b) For grimy areas, use a slightly damp cloth, then dry thoroughly
c) Check for any sticky residues that might indicate a small leak

Tip: Use this cleaning process as an opportunity to visually inspect the lines for any damage.

Troubleshooting Tip: If you notice a persistent oily residue on gas pipes, it could indicate a small leak. Perform a leak test immediately and call a professional if a leak is confirmed.

8. Maintaining Gas Detectors and Alarms

Keep your safety devices in working order:

a) Test gas and carbon monoxide detectors monthly
b) Replace batteries at least annually, or as recommended by the manufacturer
c) Vacuum or dust detectors to prevent false alarms
d) Replace entire units according to the manufacturer's recommendations, usually every 5-7 years

Tip: Use daylight saving time changes as a reminder to check and maintain your detectors.

Troubleshooting Tip: If a detector starts chirping intermittently, it usually means the batteries need replacement. If this doesn't solve the issue, the unit might be at the end of its life and need replacement.

9. Cleaning Around Gas Meters

Maintain clear access to your gas meter:

a) Keep the area around the meter free of debris and vegetation
b) Gently brush snow and ice off the meter in winter
c) Never paint over the meter or its pipes
d) Ensure the meter is protected from potential impact (e.g., from vehicles)

Tip: Trim back any bushes or plants near the meter to allow for easy access and reading.

Troubleshooting Tip: If you notice any damage to the gas meter, no matter how minor, contact your gas company immediately. Never attempt to repair a gas meter yourself.

10. Lubricating Gas Valves

Prevent valves from seizing:

a) Identify which valves can be safely maintained (usually main shut-off valves)
b) Apply a silicone-based lubricant specifically designed for gas valves
c) Turn the valve on and off several times to distribute the lubricant

Tip: Only lubricate valves that feel stiff. Over-lubrication can attract dirt and cause issues.

Troubleshooting Tip: If a valve doesn't turn smoothly after lubrication, don't force it. Call a professional to inspect and potentially replace the valve.

Remember, while regular cleaning and upkeep are important, safety should always be your top priority. If you're ever unsure about a maintenance task, or if you encounter any signs of gas leaks or malfunctions, don't hesitate to call a professional. The small cost of professional maintenance is far outweighed by the safety and peace of mind it provides.

By incorporating these cleaning and upkeep practices into your regular home maintenance routine, you're not just ensuring the efficiency of your gas system – you're actively contributing to the safety and longevity of your home. In our next section, we'll discuss emergency shut-off procedures, an essential skill for every homeowner with gas appliances. Are you ready to learn how to respond quickly and safely in a gas emergency? Let's move on to this crucial topic!

Emergency Shut-off Procedures

Imagine this scenario: It's late at night, and suddenly you smell gas in your home. Your heart races as you realize you need to act fast. In moments like these, knowing how to quickly and safely shut off your gas supply can be the difference between a close call and a potential disaster.

When I first faced a gas emergency, I felt panicked and unsure. Over time, I've learned that being prepared and knowing exactly what to do can turn a frightening situation into a manageable one. Let me guide you through the essential emergency shut-off procedures, sharing insights I've gained to help you respond confidently and safely in a gas emergency.

1. Recognizing a Gas Emergency

First, let's identify situations that require emergency shut-off:

- Strong smell of gas (rotten egg odor)
- Hissing or whistling sounds near gas lines or appliances
- Visible damage to gas pipes or appliances
- Carbon monoxide detector alarms

Tip: Trust your senses. If something seems off, it's better to be safe than sorry.

Troubleshooting Tip: If you're unsure whether you're smelling gas or something else, err on the side of caution and treat it as a gas emergency.

Tip: Have emergency contact numbers (including your gas company's) stored in your phone and posted in a visible location.

2. Immediate Actions

If you suspect a gas leak or emergency:

a) Don't panic, but act quickly
b) Do not turn on or off any electrical devices, including light switches
c) Do not use phones (including cell phones) inside the house
d) Do not light matches or create any sparks or flames
e) Open windows and doors to ventilate the area, if possible

Tip: Create a mental checklist of these steps and practice them with your family.

Troubleshooting Tip: If you're in a dark area and need light, use a battery-powered flashlight, not your phone's flashlight, to avoid potential sparks.

3. Locating Your Gas Meter

Know where your gas meter is located before an emergency occurs:

- Typically found on the exterior of your home
- May be in a basement or utility room in some homes
- Look for a metal box with pipes coming out of it

Tip: Show all family members where the gas meter is located during a non-emergency time.

Troubleshooting Tip: If you can't find your gas meter, contact your gas company for assistance. Don't wait for an emergency to locate it.

4. Shutting Off the Main Gas Valve

Here's how to turn off the main gas supply:

a) Locate the shut-off valve on the gas meter
b) Use an adjustable wrench to turn the valve 1/4 turn
c) The valve is off when the tang (the flat metal tab) is perpendicular to the pipe

Tip: Keep an adjustable wrench near your gas meter for quick access in emergencies.

Troubleshooting Tip: If the valve is difficult to turn, don't force it. Call the gas company or emergency services immediately.

5. Individual Appliance Shut-offs

For localized issues, you can shut off gas to specific appliances:

a) Locate the shut-off valve near the appliance
b) Turn the valve perpendicular to the gas line
c) Confirm the appliance is no longer receiving gas

Tip: Familiarize yourself with the location of each appliance's shut-off valve during routine maintenance.

Troubleshooting Tip: If an appliance doesn't have an easily accessible shut-off valve, consider having one installed for safety.

6. After Shutting Off the Gas

Once you've shut off the gas:

a) Evacuate the premises immediately
b) Call your gas company or emergency services from outside the building
c) Do not re-enter the building until professionals declare it safe
d) Do not attempt to turn the gas back on yourself

Troubleshooting Tip: If you're unsure whether you've successfully shut off the gas, err on the side of caution. Evacuate and let professionals handle the situation.

7. Preparing for Emergencies

Take these steps to be ready for potential gas emergencies:

a) Conduct regular family emergency drills
b) Create an emergency kit with essentials like flashlights and batteries
c) Install and maintain carbon monoxide detectors
d) Keep a wrench or gas shut-off tool near the meter

Tip: Review and update your emergency procedures annually.

Troubleshooting Tip: If family members struggle to remember emergency procedures, create a simple, visual guide and post it in a prominent place.

8. Special Considerations for Different Types of Gas

Be aware of specific procedures for different gas types:

- Natural Gas: Rises and dissipates quickly
- Propane: Heavier than air and sinks to low areas

Tip: If you use propane, pay extra attention to low-lying areas during a suspected leak.

Troubleshooting Tip: For propane systems, know the location of your tank's shut-off valve, which may be different from a natural gas meter.

9. Dealing with Outdoor Gas Leaks

For suspected outdoor leaks:

a) Move a safe distance away from the area
b) Call the gas company or emergency services immediately
c) Keep others away from the area
d) Do not attempt to locate or fix the leak yourself

Tip: Be aware of signs of outdoor leaks, like dead vegetation near gas lines or bubbling in standing water.

Troubleshooting Tip: If you smell gas outdoors but can't pinpoint the source, don't assume it's harmless. Report it to your gas company for investigation.

10. Post-Emergency Procedures

After the emergency has been resolved:

a) Wait for official clearance before re-entering your home
b) Have a professional inspect your entire gas system before use
c) Replace any emergency supplies used
d) Review and update your emergency procedures based on the experience

Tip: Use the experience as a learning opportunity to improve your emergency preparedness.

Troubleshooting Tip: If you're uncomfortable using gas appliances after an emergency, consider having a professional walk you through safe usage to rebuild your confidence.

Remember, in any gas emergency, your priority is the safety of yourself and your family. Property and possessions can be replaced, but lives cannot. When in doubt, always err on the side of caution – evacuate first and ask questions later.

By familiarizing yourself and your family with these emergency shut-off procedures, you're taking a crucial step in home safety. Practice these steps regularly so that if an emergency does occur, you can act swiftly and confidently.

In our next section, we'll discuss how to understand and improve gas appliance efficiency, helping you save money and reduce your environmental impact. Are you ready to make your gas usage more effective and economical? Let's move on to this important topic!

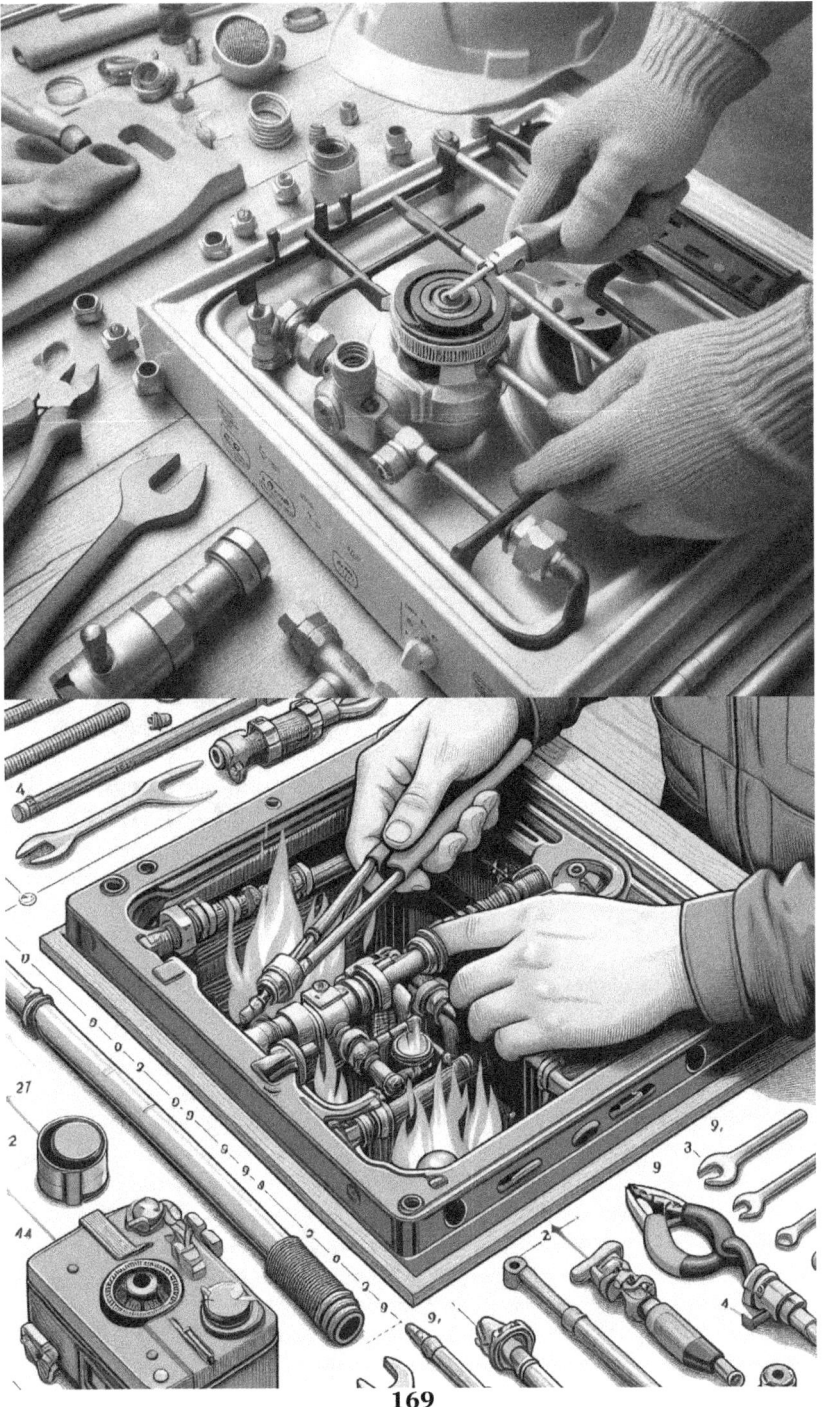

Chapter 8
Understanding Gas Appliance Efficiency

Energy Ratings Explained

Imagine standing in an appliance store, surrounded by gas-powered water heaters, furnaces, and stoves. Each boasts an energy rating, but what do these numbers and labels really mean for your home and wallet? Understanding gas appliance efficiency and energy ratings can be the key to making smart choices that save you money and reduce your environmental impact.

When I first encountered energy ratings, I found them confusing and overwhelming. Over time, I've learned that these ratings are valuable tools for comparing appliances and estimating long-term costs. Let me guide you through the world of gas appliance efficiency ratings, sharing insights I've gained to help you make informed decisions about your home's gas-powered equipment.

1. The Importance of Energy Ratings

Energy ratings serve several purposes:
- Help consumers compare the efficiency of different appliances
- Estimate potential energy costs and savings
- Encourage manufacturers to produce more efficient appliances
- Support energy conservation efforts

Tip: Always consider energy ratings when purchasing new gas appliances, even if the upfront cost is higher for more efficient models.

Troubleshooting Tip: If an appliance's actual energy use doesn't match its rating, check for installation issues or maintenance needs that could be affecting its efficiency.

2. Annual Fuel Utilization Efficiency (AFUE)

AFUE is used for gas furnaces and boilers:
- Measures the percentage of fuel converted to heat over a typical year
- Higher percentages indicate greater efficiency
- Modern high-efficiency furnaces can have AFUE ratings of 90-98%

Example: A furnace with 95% AFUE converts 95% of its fuel into heat, with only 5% lost.

Tip: Look for furnaces with AFUE ratings of 90% or higher for significant energy savings.

Troubleshooting Tip: If your furnace's efficiency seems to drop, check for dirty filters, blocked vents, or the need for professional maintenance.

3. Energy Factor (EF) for Water Heaters

EF ratings for water heaters consider:
- Efficiency of heat transfer from fuel to water
- Standby losses (heat lost from stored water)
- Cycling losses (energy lost as water circulates through a water heater)

Higher EF numbers indicate greater efficiency. For gas water heaters:
- Standard models: 0.50 to 0.60 EF
- High-efficiency models: 0.80 EF or higher

Tip: Consider tankless water heaters for even higher efficiency, with EF ratings up to 0.98.

Troubleshooting Tip: If your water heater's efficiency declines, check for sediment buildup in the tank and consider flushing it.

4. Seasonal Energy Efficiency Ratio (SEER) for Heat Pumps

While primarily used for electric heat pumps, some gas-powered heat pumps use SEER:
- Measures cooling efficiency over a typical cooling season
- Higher SEER ratings indicate greater efficiency
- Modern efficient units have SEER ratings of 18 or higher

Tip: In moderate climates, a high-SEER heat pump can be more efficient than a traditional furnace and air conditioner combination.

Troubleshooting Tip: If your heat pump's efficiency drops, check for dirty coils or low refrigerant levels.

5. Energy Star Certification

Energy Star is a voluntary program that certifies highly efficient appliances:
- Look for the blue Energy Star logo
- Certified appliances typically use 10-50% less energy than standard models
- Covers a wide range of gas appliances including furnaces, water heaters, and fireplaces

Tip: Energy Star certification can be a quick way to identify efficient appliances without diving into specific ratings.

Troubleshooting Tip: If an Energy Star appliance isn't delivering expected savings, check if it's properly sized for your home and usage patterns.

6. BTU Input and Output Ratings

BTU (British Thermal Unit) ratings are crucial for understanding an appliance's capacity:
- Input BTU: Amount of gas the appliance consumes
- Output BTU: Amount of usable heat produced

The difference between input and output reflects the appliance's efficiency.

Tip: Choose an appliance with BTU output that matches your needs – oversized units can be less efficient.

Troubleshooting Tip: If an appliance seems underpowered, check that gas pressure and burner settings are correct before assuming the BTU rating is inaccurate.

7. Thermal Efficiency for Commercial Appliances

Used for larger commercial gas appliances:
- Measures the percentage of heat transferred to the heated medium (e.g., water, air)
- Does not account for standby or cycling losses
- Higher percentages indicate greater efficiency

Tip: Even in residential settings, thermal efficiency can be useful for comparing high-capacity water heaters or boilers.

Troubleshooting Tip: If thermal efficiency drops in a commercial appliance, check for scale buildup, burner issues, or heat exchanger problems.

8. Understanding EnergyGuide Labels

Yellow EnergyGuide labels provide valuable information:
- Estimated yearly energy consumption
- Estimated yearly operating cost
- Energy consumption range for similar models

Tip: Use these labels to compare different models and estimate long-term costs.

Troubleshooting Tip: If your actual energy costs are much higher than the label estimate, check your gas rates and usage patterns, and consider a professional energy audit.

9. Combustion Efficiency

This measures how effectively an appliance burns its fuel:
- Higher percentages indicate more complete combustion
- Modern high-efficiency appliances can achieve combustion efficiency over 95%

Tip: Regular maintenance can help maintain high combustion efficiency.

Troubleshooting Tip: If you notice yellow flames instead of blue in your gas appliance, it may indicate poor combustion efficiency and require professional adjustment.

10. Interpreting Efficiency Ratings in Context

Remember that efficiency ratings are based on standardized tests:
- Actual performance can vary based on installation, maintenance, and usage patterns
- Consider your specific needs and local energy costs when interpreting ratings

Tip: Use online calculators to estimate potential savings based on your local gas rates and usage patterns.

Troubleshooting Tip: If you're not seeing expected savings from a high-efficiency appliance, consider having a professional assess your entire system for optimization opportunities.

Understanding energy ratings empowers you to make informed decisions about your gas appliances. While higher-efficiency models often cost more upfront, they can lead to significant savings over the appliance's lifetime. Moreover, choosing efficient appliances reduces your carbon footprint, contributing to broader energy conservation efforts.

Remember, efficiency ratings are just one part of the equation. Proper installation, regular maintenance, and smart usage habits all play crucial roles in realizing the full potential of your gas appliances' efficiency.

In our next section, we'll explore strategies for upgrading to high-efficiency models, helping you navigate the process of improving your home's overall energy performance. Are you ready to take your gas appliance efficiency to the next level? Let's move on to this exciting topic!

Upgrading to High-Efficiency Models

Imagine your energy bills steadily decreasing while your home comfort increases. This scenario isn't just a dream—it's entirely possible by upgrading to high-efficiency gas appliances. But the process of upgrading can seem daunting. Where do you start? How do you choose? What should you expect?

When I first considered upgrading my gas appliances, I felt overwhelmed by the options and uncertain about the potential benefits. Over time, I've learned that upgrading to high-efficiency models can be a game-changer for both your wallet and your home's comfort. Let me guide you through the process of upgrading to high-efficiency gas appliances, sharing insights I've gained to help you make smart, cost-effective choices.

1. Assessing Your Current Situation

Before upgrading, evaluate your existing setup:

a) Age of current appliances
b) Energy bills over the past year
c) Comfort levels in your home
d) Any recurring issues with existing appliances

Tip: Create a spreadsheet to track this information, making it easier to compare potential savings later.

Troubleshooting Tip: If your energy bills seem unusually high, have an energy audit done before upgrading. There might be other issues, like poor insulation, affecting your energy use.

Tip: Give yourself time to learn and adjust to your new appliance. It might take a few months to see the full benefits.

2. Identifying Upgrade Opportunities

Focus on appliances that will give you the biggest bang for your buck:

a) Furnace or boiler (if over 15 years old)
b) Water heater (if over 10 years old)
c) Gas fireplace insert (to replace an inefficient wood-burning fireplace)
d) Gas stove or oven (if very old or inefficient)

Tip: Prioritize appliances you use most frequently for maximum impact on your energy bills.

Troubleshooting Tip: If an appliance is relatively new but inefficient, weigh the cost of upgrading against potential savings. Sometimes, improving insulation or sealing air leaks can be more cost-effective.

3. Understanding High-Efficiency Features

Look for these features in high-efficiency gas appliances:

a) Condensing technology (for furnaces and water heaters)
b) Modulating gas valves
c) Variable-speed blowers
d) Electronic ignition (instead of standing pilot lights)
e) Programmable or smart controls

Tip: Research these features to understand how they contribute to efficiency. Knowledge is power when making your selection.

Troubleshooting Tip: If a high-efficiency feature seems too complex, don't shy away. Many come with user-friendly interfaces and can significantly boost your savings.

4. Calculating Potential Savings

Estimate your return on investment:

a) Use online calculators provided by Energy Star or manufacturers
b) Consider local gas rates and your typical usage
c) Factor in any available rebates or tax incentives

Tip: Be conservative in your estimates to avoid disappointment. It's better to be pleasantly surprised by higher savings.

Troubleshooting Tip: If calculated savings seem too good to be true, double-check your inputs and assumptions. Overly optimistic calculations can lead to unrealistic expectations.

5. Sizing Your New Appliance

Ensure your new appliance is properly sized for your home:

a) Have a professional perform a load calculation
b) Consider any recent home improvements (like added insulation) that might affect sizing
c) Don't automatically replace with the same size as your old appliance

Tip: An oversized appliance can be just as inefficient as an undersized one. Proper sizing is crucial for optimal efficiency.

Troubleshooting Tip: If you're getting conflicting size recommendations, ask each professional to explain their calculation method. Understanding their reasoning can help you make an informed decision.

6. Choosing the Right Model

Consider these factors when selecting your high-efficiency appliance:

a) Energy Star certification
b) Highest efficiency rating in your budget
c) Brand reputation and warranty
d) Additional features that might enhance comfort or convenience

Tip: Read user reviews, but take them with a grain of salt. Focus on reviews that discuss long-term performance and efficiency.

Troubleshooting Tip: If you're torn between two models, consider the availability of local service technicians familiar with each brand. Good maintenance support can be as important as initial efficiency.

7. Planning for Installation

Proper installation is crucial for high-efficiency appliances:

a) Choose a certified professional installer
b) Ensure your home can accommodate new venting requirements (especially for condensing appliances)
c) Consider any electrical upgrades needed for new controls or blowers
d) Plan for proper drainage of condensate (for condensing appliances)

Tip: Get multiple installation quotes and ask each installer about their experience with high-efficiency models.

Troubleshooting Tip: If an installer suggests shortcuts that don't align with manufacturer recommendations, get a second opinion. Proper installation is key to achieving stated efficiency levels.

8. Understanding New Maintenance Requirements

High-efficiency appliances may have different maintenance needs:

a) Familiarize yourself with the maintenance schedule
b) Budget for professional annual maintenance
c) Learn which tasks you can do yourself (like changing filters) and which require a pro

Tip: Set reminders for regular maintenance tasks to ensure your new appliance continues operating at peak efficiency.

Troubleshooting Tip: If you notice any decrease in performance or increase in energy bills, don't wait for your annual check-up. Address issues promptly to maintain efficiency.

9. Maximizing Your Upgrade

Get the most out of your new high-efficiency appliance:

a) Adjust your habits to take advantage of new features (like programming your thermostat)
b) Consider complementary upgrades (like improved insulation or smart home integration)
c) Monitor your energy bills to track actual savings

Troubleshooting Tip: If you're not seeing expected savings after a few months, consult with your installer or the manufacturer. There might be settings that need adjustment or usage patterns you can optimize.

10. Disposing of Your Old Appliance

Responsibly dispose of your old gas appliance:

a) Check if your installer offers disposal services
b) Look for local recycling programs that accept gas appliances
c) Ensure any refrigerants or hazardous materials are properly handled

Tip: Some utilities offer "bounty" programs that pay you for removing old, inefficient appliances from the grid.

Troubleshooting Tip: If you're considering keeping your old appliance as a backup, weigh this against the energy cost of running an inefficient unit. It's often better to fully commit to your upgrade.

Upgrading to high-efficiency gas appliances is a significant step towards a more comfortable, cost-effective, and environmentally friendly home. While the process may seem complex, the long-term benefits in reduced energy bills, improved comfort, and lower environmental impact make it a worthwhile investment.

Remember, every home is unique, and what works best for one might not be ideal for another. Take your time, do your research, and don't hesitate to consult with professionals. The goal is not just to upgrade your appliances, but to upgrade your entire home experience.

In our next section, we'll explore practical tips for saving energy with your gas appliances, helping you maximize the efficiency of both new and existing equipment. Are you ready to become a master of gas energy conservation? Let's move on to these valuable strategies!

Cost-Saving Tips

Imagine opening your gas bill and smiling instead of wincing. This isn't a fantasy—it's entirely possible with smart, cost-saving strategies for your gas appliances. Whether you've just upgraded to high-efficiency models or you're working with older equipment, there are always ways to trim your gas consumption and save money.

When I first started looking for ways to reduce my gas bills, I was surprised by how many simple, effective strategies were available. Over time, I've learned that a combination of smart habits, minor adjustments, and strategic investments can lead to significant savings. Let me guide you through a range of cost-saving tips for your gas appliances, sharing insights I've gained to help you maximize efficiency and minimize costs.

1. Optimizing Your Thermostat

Your thermostat is the control center for your heating costs:

a) In winter, set your thermostat to 68°F (20°C) when awake and lower when asleep or away
b) In summer, if you have gas-powered AC, set it to 78°F (26°C) when home
c) Install a programmable or smart thermostat to automate these changes

Tip: Each degree you lower your thermostat in winter can save up to 3% on your heating bill.

Troubleshooting Tip: If your programmable thermostat isn't saving you money, check its location. It should be away from drafts, sunlight, and heat sources that could affect its readings.

2. Water Heater Efficiency

Your water heater can be a major gas consumer:

a) Lower the temperature to 120°F (49°C) for both safety and efficiency
b) Insulate your water heater tank and hot water pipes
c) Use low-flow fixtures to reduce hot water consumption
d) Drain a quart of water from the tank every three months to remove sediment

Tip: Consider installing a timer on your water heater to turn it off when you typically don't use hot water.

Troubleshooting Tip: If you're running out of hot water frequently, your water heater might be undersized or need maintenance, rather than needing a higher temperature setting.

3. Efficient Cooking Practices

Make your gas stove work smarter, not harder:

a) Use lids on pots and pans to cook food more quickly
b) Match pot size to burner size for optimal heat transfer
c) Use a kettle or electric water heater for boiling water, it's often more efficient
d) Keep burners clean for better flame efficiency

Tip: Consider using a pressure cooker or slow cooker for long-cooking dishes, as they're often more efficient than stovetop cooking.

Troubleshooting Tip: If your gas burners produce a yellow flame instead of blue, they need cleaning or adjustment. A yellow flame is less efficient and can produce harmful carbon monoxide.

4. Furnace and Boiler Maintenance

Regular maintenance keeps your heating system efficient:

a) Replace or clean filters monthly during heavy-use seasons
b) Have your system professionally serviced annually
c) Keep vents and radiators clean and unobstructed
d) Seal and insulate ductwork to prevent heat loss

Tip: Consider installing a furnace whistle, which alerts you when it's time to change the filter.

Troubleshooting Tip: If some rooms are much colder than others, your system might be unbalanced. Have a professional check and adjust your system's balance for even, efficient heating.

5. Fireplace Efficiency

Gas fireplaces can be cozy, but also costly if not used efficiently:

a) Use your fireplace for zone heating, lowering the thermostat in the rest of the house
b) Install glass doors to reduce heat loss up the chimney
c) Turn off the pilot light in summer months
d) Consider a high-efficiency fireplace insert if you use your fireplace frequently

Tip: If you rarely use your fireplace, consider installing an inflatable chimney balloon to prevent heat loss when it's not in use.

Troubleshooting Tip: If your gas fireplace isn't providing much heat, check the fan (if it has one). A malfunctioning fan can significantly reduce a fireplace's heating efficiency.

6. Sealing and Insulation

Prevent your gas-heated air from escaping:

a) Seal air leaks around windows, doors, and other openings
b) Add insulation to your attic, walls, and floors
c) Install storm windows or energy-efficient replacements
d) Use heavy curtains or cellular shades to reduce heat loss through windows

Tip: A simple way to check for air leaks is to hold a lit incense stick near potential leak areas on a windy day. If the smoke wavers, you've found a leak.

Troubleshooting Tip: If you've added insulation but aren't seeing expected savings, check for missed areas or compressed insulation, which reduces its effectiveness.

7. Smart Appliance Use

Use your gas appliances strategically:

a) Run full loads in the dishwasher and washing machine
b) Use cold water for laundry when possible
c) Take shorter showers to reduce hot water use
d) Consider gas-powered outdoor appliances (like grills) for summer cooking to keep heat out of the house

Tip: If you have a gas dryer, clean the lint filter before every load to maintain efficiency.

Troubleshooting Tip: If your clothes aren't drying efficiently, check the dryer vent for blockages. A clogged vent makes your dryer work harder and use more gas.

8. Zoned Heating

Heat only the areas you're using:

a) Close vents and doors in unused rooms
b) Use space heaters judiciously for small areas instead of heating the whole house
c) Consider installing a zoned heating system for more precise control

Tip: Remember to keep some air flowing to all areas to prevent issues like frozen pipes in very cold weather.

Troubleshooting Tip: If closing vents causes whistling or uneven heating, your system might not be designed for zoning. Consult with an HVAC professional about the best zoning strategy for your setup.

9. Seasonal Adjustments

Adapt your habits to the seasons:

a) In summer, use natural gas appliances early in the morning or late at night to minimize indoor heat gain
b) In winter, open curtains during the day to let in sunlight and close them at night to retain heat
c) Adjust your water heater temperature seasonally if your usage patterns change

Tip: Consider a gas-powered clothes drying rack for winter use. It can dry clothes efficiently while adding heat and humidity to your home.

Troubleshooting Tip: If your gas bills spike dramatically in one season, compare year-over-year usage. A sudden increase could indicate an efficiency problem with your heating or cooling system.

10. Monitor and Analyze Your Usage

Keep track of your gas consumption:

a) Read your gas meter regularly and track your usage
b) Compare your bills year-over-year, accounting for temperature differences
c) Use energy monitoring tools or smart home systems to get detailed usage data

Tip: Many utilities offer online tools to help you analyze your energy usage patterns.

Troubleshooting Tip: If you notice a sudden increase in gas usage without a clear cause, check for leaks or malfunctioning appliances immediately.

Remember, saving money on your gas bill is not just about reducing comfort—it's about using energy more intelligently. Many of these tips can actually increase your comfort while decreasing your bills. Start with the easiest changes and work your way up to more significant adjustments or investments.

Also, keep in mind that what works best can vary based on your specific home, climate, and lifestyle. Don't be afraid to experiment with different strategies to find what gives you the best balance of comfort and savings.

By implementing these cost-saving tips, you're not just reducing your gas bills—you're also decreasing your carbon footprint and contributing to energy conservation efforts. It's a win-win for your wallet and the environment.

In our next section, we'll explore advanced gas line projects for those looking to expand their home's gas capabilities. Are you ready to take on some more complex gas-related home improvements? Let's move on to these exciting possibilities!

Chapter 9
Advanced Gas Line Projects

Installing Gas Lighting

Imagine stepping into your backyard on a cool evening, greeted by the warm, flickering glow of gas lanterns lining your patio or walkway. Gas lighting can add a touch of classic elegance and ambiance to your outdoor spaces, while also providing reliable illumination. However, installing gas lighting is a more complex project that requires careful planning and execution.

When I first considered adding gas lighting to my home, I was drawn to its unique charm but intimidated by the technical aspects of the installation. Over time, I've learned that while this project is more advanced, it's achievable with the right knowledge and precautions. Let me guide you through the process of installing gas lighting, sharing insights I've gained to help you tackle this rewarding project safely and effectively.

1. Planning Your Gas Lighting Project

Before you begin, carefully plan your installation:

a) Determine the locations for your gas lights
b) Decide on the style and number of fixtures
c) Assess your existing gas line capacity
d) Check local codes and permit requirements

Tip: Create a detailed sketch of your property, marking proposed light locations and existing gas lines.

Troubleshooting Tip: If you're unsure about your gas system's capacity, consult a professional. Adding too many lights could strain your system and affect other appliances.

2. Selecting Gas Light Fixtures

Choose fixtures that suit your needs and style:

a) Consider brightness levels (measured in lumens or BTUs)
b) Look for weather-resistant materials for outdoor use
c) Decide between manual, automatic, or smart ignition systems
d) Check for safety certifications (like CSA or UL listings)

Tip: Opt for fixtures with easy-to-replace mantles and clear access for maintenance.

Troubleshooting Tip: If a fixture seems dimmer than expected after installation, check the gas pressure and orifice size. They may need adjustment for optimal performance.

3. Sizing and Running New Gas Lines

Extend your existing gas system to reach your new lights:

a) Calculate the total gas load, including existing appliances
b) Determine the appropriate pipe size based on distance and load
c) Plan the most direct and protected route for new gas lines
d) Consider using flexible gas lines for easier installation in tight spaces

Tip: Always overestimate your gas needs slightly to allow for future expansion.

Troubleshooting Tip: If you encounter unexpected obstacles while running lines, don't force a path. Reassess your route to avoid potential hazards or structural issues.

4. Trenching and Pipe Installation

For outdoor lighting, you'll likely need to bury gas lines:

a) Call utility companies to mark existing underground lines
b) Dig trenches at least 18 inches deep (or to local code requirements)
c) Use appropriate outdoor-rated gas piping
d) Include a tracer wire with plastic pipes for future detection

Tip: Slightly slope long horizontal runs back towards the gas source to prevent condensation buildup.

Troubleshooting Tip: If you hit rock or tree roots while digging, don't try to force through. You may need to adjust your path or use specialized equipment.

5. Installing Shut-Off Valves

Include shut-off valves for safety and convenience:

a) Install a main shut-off valve where the new line branches from your existing system
b) Place individual shut-off valves near each light fixture
c) Ensure all valves are easily accessible

Tip: Use quarter-turn ball valves for easy operation and reliable sealing.

Troubleshooting Tip: If a valve feels stiff or doesn't turn smoothly, don't force it. Replace it immediately to prevent potential leaks or failures.

6. Mounting Light Fixtures

Securely install your gas light fixtures:

a) Follow manufacturer instructions for mounting
b) Ensure fixtures are level and properly aligned
c) Use appropriate weatherproof sealants around mounting points
d) Connect the gas line, using appropriate fittings and sealants

Tip: If mounting on masonry, use anchors designed for gas fixture installation to ensure stability.

Troubleshooting Tip: If a fixture wobbles after installation, don't ignore it. This could lead to gas line stress and potential leaks. Remount or reinforce as necessary.

7. Testing the System

Thoroughly test your new gas lighting system:

a) Conduct a pressure test on all new gas lines before connecting fixtures
b) Check all connections with a gas leak detector or soapy water solution
c) Test each fixture individually, checking for proper ignition and flame quality
d) Observe the system over several days to ensure consistent performance

Tip: Perform tests at different times of day and in various weather conditions to ensure reliability.

Troubleshooting Tip: If you detect any leaks, no matter how small, shut off the gas immediately and repair before proceeding. Never ignore or try to temporarily patch a gas leak.

8. Adjusting for Optimal Performance

Fine-tune your gas lights for the best results:

a) Adjust gas pressure if needed for proper flame height and brightness
b) Clean and align mantles for even illumination
c) Set timers or photocells for automatic operation, if applicable
d) Balance gas flow if you have multiple lights on the same line

Tip: Keep a log of any adjustments you make for future reference.

Troubleshooting Tip: If lights flicker or dim unexpectedly, check for drafts or obstructions near the fixture. You may need to install wind guards or relocate the light.

9. Safety Considerations

Prioritize safety throughout the installation and use:

a) Ensure proper ventilation for all gas lights, especially if any are installed in covered areas
b) Install carbon monoxide detectors in adjacent indoor spaces
c) Create clear instructions for operation and emergency shut-off procedures
d) Schedule regular professional inspections of your gas lighting system

Tip: Consider installing an automatic shut-off system that cuts gas flow in case of earthquakes or other emergencies.

Troubleshooting Tip: If you ever smell gas near your lighting fixtures, turn off the gas supply immediately and call a professional. Don't attempt to locate or repair the leak yourself.

10. Maintenance and Upkeep

Regular maintenance keeps your gas lights safe and efficient:

a) Clean fixtures regularly to prevent debris buildup
b) Replace mantles as needed (typically every 6-12 months)
c) Check and clean burner orifices annually
d) Inspect gas lines and connections for signs of wear or damage

Tip: Create a maintenance schedule and set reminders to ensure regular upkeep.

Troubleshooting Tip: If a light stops working suddenly, don't assume it's a major issue. Often, it's just a clogged orifice or worn-out mantle that's easily fixed.

Installing gas lighting is a project that requires careful planning, precise execution, and a strong commitment to safety. While it's more complex than many DIY tasks, the result can be a stunning addition to your home that provides both practical illumination and aesthetic appeal.

Remember, if at any point you feel uncomfortable or unsure about any aspect of the installation, don't hesitate to consult with or hire a professional. The safety of your home and family should always be the top priority.

By successfully installing gas lighting, you're not just adding a beautiful feature to your home – you're also expanding your skills and knowledge in working with gas systems. This experience can be valuable for future projects and home maintenance.

In our next section, we'll explore another advanced gas line project: connecting a pool heater. Are you ready to learn how to extend your swimming season with a gas-powered pool heater? Let's dive into this exciting topic!

Pool Heater Connections

Imagine extending your swimming season well into the cooler months, or taking a comfortable dip on a chilly summer evening. A gas pool heater can make this dream a reality, providing efficient and powerful heating for your pool. However, connecting a pool heater to your gas line is a complex project that requires careful planning and execution.

When I first considered adding a gas pool heater, I was excited about the possibilities but also aware of the challenges involved. Over time, I've learned that while this project is more advanced, it's achievable with the right approach and attention to detail. Let me guide you through the process of connecting a pool heater to your gas line, sharing insights I've gained to help you tackle this project safely and effectively.

1. Planning Your Pool Heater Installation

Before you begin, carefully plan your project:

a) Determine the ideal location for your pool heater
b) Calculate the appropriate heater size for your pool
c) Assess your existing gas line capacity
d) Check local codes and permit requirements

Tip: Choose a location that's close to both your pool equipment and gas supply to minimize pipe runs.

Troubleshooting Tip: If your existing gas line can't support the heater's needs, you may need to upgrade your gas meter or line size. Consult with your gas company early in the planning process.

Tip: Consider installing a pool cover to retain heat and reduce heating costs when the pool isn't in use.

2. Selecting the Right Pool Heater

Choose a heater that suits your needs:

a) Consider the BTU output required for your pool size and desired temperature rise
b) Look for energy-efficient models with high thermal efficiency ratings
c) Decide between millivolt and electronic ignition systems
d) Check for safety features like auto shut-off and diagnostic systems

Tip: Opt for a heater with a digital display for easy temperature control and troubleshooting.

Troubleshooting Tip: If your heater seems undersized after installation, double-check your calculations. Factors like wind exposure and humidity can affect heating needs.

3. Preparing the Installation Site

Create a stable and suitable base for your heater:

a) Install a level concrete pad that's slightly elevated to prevent water accumulation
b) Ensure proper clearances around the heater for ventilation and maintenance
c) Consider sound dampening if the heater is near living areas
d) Plan for proper drainage of condensation and rainwater

Tip: Make the concrete pad a few inches larger than the heater's footprint to allow for easy access.

Troubleshooting Tip: If you notice excessive vibration after installation, check the levelness of your pad and the heater's mounting. Uneven surfaces can cause vibration and potential damage.

4. Running the Gas Line

Extend your gas system to reach the pool heater:

a) Calculate the correct pipe size based on the heater's BTU rating and distance from the gas meter
b) Use appropriate outdoor-rated gas piping
c) Install a sediment trap before the heater connection
d) Include a properly sized and easily accessible shut-off valve

Tip: Always oversize your gas line slightly to allow for future expansion or increased demand.

Troubleshooting Tip: If you experience low gas pressure at the heater, check for any kinks or unnecessary bends in the line. Long runs or too many fittings can reduce pressure.

5. Venting Considerations

Proper venting is crucial for safe operation:

a) Follow manufacturer guidelines for vent size and type
b) Ensure the vent terminates at a safe distance from windows, doors, and air intakes
c) Use corrosion-resistant materials for outdoor venting
d) Install a rain cap to prevent water entry

Tip: Consider wind patterns when positioning the vent termination to prevent exhaust from blowing towards your pool or outdoor living areas.

Troubleshooting Tip: If you notice unusual odors or poor heater performance, check the venting system for blockages or damage. Birds' nests or debris can obstruct proper ventilation.

6. Electrical Connections

Most gas pool heaters require some electrical work:

a) Ensure a GFCI-protected outlet is available near the heater
b) Follow manufacturer guidelines for voltage and amperage requirements
c) Use waterproof connections and conduit for all outdoor wiring
d) Bond the heater to your pool's bonding grid

Tip: Consider installing a separate circuit for your pool heater to prevent overloading existing circuits.

Troubleshooting Tip: If the heater won't start, check both the gas and electrical connections. Sometimes, a tripped GFCI can be mistaken for a gas supply issue.

7. Connecting to Pool Plumbing

Integrate the heater into your pool's circulation system:

a) Install bypass valves to allow for heater isolation when not in use
b) Use sweep elbows instead of 90-degree fittings to reduce flow restriction
c) Ensure proper flow rate through the heater as per manufacturer specifications

d) Install a check valve to prevent backflow if the heater is below water level

Tip: Use union fittings at the heater connections for easier removal and maintenance.

Troubleshooting Tip: If you notice reduced flow after installation, check for air in the system. Improper plumbing can introduce air, reducing efficiency and potentially damaging the heater.

8. Testing the System

Thoroughly test your new pool heating system:

a) Conduct a pressure test on the gas line before connecting the heater
b) Check all gas connections with a leak detector or soapy water solution
c) Purge air from the water lines before starting the heater
d) Test the heater through a complete heating cycle

Tip: Monitor the system closely during the first few days of operation, checking for any unusual noises, smells, or performance issues.

Troubleshooting Tip: If the heater short-cycles (turns on and off frequently), check the flow rate. Insufficient water flow can cause the heater to shut off prematurely.

9. Safety Features and Considerations

Prioritize safety in your installation:

a) Install carbon monoxide detectors in adjacent enclosed spaces
b) Ensure all required safety switches (like pressure and temperature limits) are functioning
c) Post clear operating instructions and emergency shut-off procedures near the heater
d) Keep the area around the heater clear of flammable materials

Troubleshooting Tip: If any safety feature activates repeatedly, don't override it. This usually indicates a problem that needs professional attention.

10. Maintenance and Upkeep

Regular maintenance keeps your pool heater safe and efficient:

a) Clean or replace filters regularly to ensure proper water flow
b) Inspect the heat exchanger annually for scale buildup or corrosion
c) Check gas connections and venting system for any signs of wear or damage
d) Test safety features at the start of each swimming season

Tip: Create a maintenance log to track service dates, repairs, and any adjustments made.

Troubleshooting Tip: If heating efficiency decreases over time, check for scale buildup in the heat exchanger. In hard water areas, this can significantly reduce performance.

Connecting a gas pool heater is a complex project that requires a good understanding of gas systems, plumbing, and electrical work. While it's more challenging than many DIY tasks, the reward is a significantly extended swimming season and more comfortable pool use.

Remember, if at any point you feel unsure about any aspect of the installation, don't hesitate to consult with or hire a professional. The safety of your family and the protection of your investment should always be the top priorities.

By successfully connecting a gas pool heater, you're not just enhancing your pool experience – you're also expanding your skills in working with gas systems and pool equipment. This knowledge can be valuable for future projects and ongoing maintenance of your pool system.

Whole-House Generator Hookups

Imagine this scenario: A severe storm has knocked out power in your neighborhood. While your neighbors sit in darkness, your home hums with the reassuring sound of a whole-house generator, keeping your lights on, your food cold, and your family comfortable. A whole-house generator can provide peace of mind and continuity during power outages, but connecting one to your gas line is a complex project that requires careful planning and execution.

When I first considered installing a whole-house generator, I was excited about the prospect of energy independence but also aware of the technical challenges involved. Over time, I've learned that while this project is advanced, it's achievable with the right approach and attention to detail. Let me guide you through the process of connecting a whole-house generator to your gas line, sharing insights I've gained to help you tackle this project safely and effectively.

1. Planning Your Generator Installation

Before you begin, carefully plan your project:

a) Determine the appropriate generator size based on your home's power needs
b) Choose an ideal location for the generator (outdoors, away from windows and doors)
c) Assess your existing gas line capacity
d) Check local codes, permits, and potential noise ordinances

Tip: Create a list of essential appliances and systems you want to power during an outage to help size your generator correctly.

Troubleshooting Tip: If your existing gas line can't support the generator's needs, you may need to upgrade your gas meter or line size. Consult with your gas company early in the planning process.

2. Selecting the Right Generator

Choose a generator that suits your needs:

a) Consider the kilowatt (kW) output required for your home
b) Look for models with automatic transfer switches for seamless power transition
c) Decide between air-cooled and liquid-cooled systems
d) Check for features like remote monitoring and self-testing capabilities

Tip: Opt for a generator with a low-speed engine (1800 rpm) for quieter operation and longer life.

Troubleshooting Tip: If your generator seems to struggle under load after installation, double-check your power calculations. You may have underestimated your needs or forgotten to account for starting wattages of some appliances.

3. Preparing the Installation Site

Create a stable and suitable base for your generator:

a) Install a level concrete pad that's slightly elevated to prevent water accumulation
b) Ensure proper clearances around the generator for ventilation and maintenance
c) Consider sound dampening measures, like acoustic fencing
d) Plan for proper drainage of rainwater and engine fluids

Tip: Make the concrete pad at least 4 inches thick and extend it 6 inches beyond the generator's footprint on all sides.

Troubleshooting Tip: If you notice excessive vibration after installation, check the levelness of your pad and the generator's mounting. Uneven surfaces can cause vibration and potential damage.

4. Running the Gas Line

Extend your gas system to reach the generator:

a) Calculate the correct pipe size based on the generator's BTU rating and distance from the gas meter
b) Use appropriate outdoor-rated gas piping
c) Install a sediment trap before the generator connection
d) Include a properly sized and easily accessible shut-off valve

Tip: Always oversize your gas line slightly to allow for the generator's peak fuel consumption during startup.

Troubleshooting Tip: If you experience low gas pressure at the generator, check for any kinks or unnecessary bends in the line. Long runs or too many fittings can reduce pressure.

5. Electrical Connections

Connecting a whole-house generator requires significant electrical work:

a) Install a transfer switch to safely disconnect from the grid during outages
b) Connect the generator to your home's electrical panel
c) Ensure all connections are weatherproof and up to code

d) Consider installing a power management system to prioritize circuits

Tip: Always hire a licensed electrician for this part of the installation. Improper electrical connections can be extremely dangerous.

Troubleshooting Tip: If your generator starts but doesn't power your home, check the transfer switch connections. A faulty transfer switch can prevent power from reaching your home's circuits.

6. Ventilation and Exhaust Considerations

Proper ventilation is crucial for safe operation:

a) Ensure the generator is placed where exhaust can't enter the home
b) Direct exhaust away from neighbors' properties
c) Consider prevailing wind directions when positioning the exhaust
d) Install carbon monoxide detectors in your home as an extra precaution

Tip: Use a CO detector with a digital readout to monitor CO levels precisely.

Troubleshooting Tip: If you smell exhaust in your home during generator operation, shut it down immediately and recheck all seals and ventilation paths. Even small leaks can be dangerous.

7. Fuel System Setup

Ensure a reliable fuel supply for your generator:

a) Install a dedicated gas line from your meter to the generator
b) Use a flexible connector at the generator to absorb vibrations
c) Install a gas pressure regulator if required by the generator manufacturer
d) Consider a backup propane tank for extended outages if your generator is dual-fuel capable

Tip: Install a gas flow meter to monitor your generator's fuel consumption over time.

Troubleshooting Tip: If your generator runs rough or inconsistently, check the gas pressure. Fluctuating or incorrect gas pressure can cause performance issues.

8. Control System Installation

Set up the generator's control system:

a) Install the automatic transfer switch according to manufacturer instructions
b) Set up any remote monitoring systems or apps
c) Program the generator's exercise schedule to run weekly
d) Configure any power management modules to prioritize essential circuits

7. Fuel System Setup

Ensure a reliable fuel supply for your generator:

a) Install a dedicated gas line from your meter to the generator
b) Use a flexible connector at the generator to absorb vibrations
c) Install a gas pressure regulator if required by the generator manufacturer
d) Consider a backup propane tank for extended outages if your generator is dual-fuel capable

Tip: Install a gas flow meter to monitor your generator's fuel consumption over time.

Troubleshooting Tip: If your generator runs rough or inconsistently, check the gas pressure. Fluctuating or incorrect gas pressure can cause performance issues.

8. Control System Installation

Set up the generator's control system:

a) Install the automatic transfer switch according to manufacturer instructions
b) Set up any remote monitoring systems or apps
c) Program the generator's exercise schedule to run weekly
d) Configure any power management modules to prioritize essential circuits

Tip: Place the generator's control panel in an easily accessible location for quick monitoring and manual operation if needed.

Troubleshooting Tip: If your generator fails to start automatically during an outage, check the control system's settings and connections. Sometimes, a loose wire or incorrect setting can prevent auto-start.

9. Testing the System

Thoroughly test your new whole-house generator system:

a) Conduct a pressure test on the gas line before connecting the generator
b) Check all gas connections with a leak detector or soapy water solution
c) Perform a full-load test to ensure the generator can handle your home's power needs
d) Test the automatic transfer switch under various scenarios

Tip: Simulate a power outage by shutting off your main breaker to ensure a smooth transition to generator power.

Troubleshooting Tip: If your generator powers on but some circuits don't work, check your transfer switch and power management system settings. You may need to adjust load shedding priorities.

10. Maintenance and Regular Testing

Establish a routine to keep your generator ready for action:

a) Schedule annual professional maintenance
b) Perform weekly self-tests to ensure the system is operational

c) Check and change oil according to the manufacturer's schedule

d) Keep a maintenance log to track service dates and any issues

Tip: Consider a maintenance contract with a local generator service company for regular check-ups and priority service during outages.

Troubleshooting Tip: If your generator fails to perform its weekly self-test, check the battery. A weak or dead battery is a common cause of start failures.

Installing a whole-house generator is a complex project that requires expertise in gas systems, electrical work, and mechanical installation. While it's one of the most challenging home improvement projects, the peace of mind and comfort it provides during power outages can be invaluable.

Remember, if at any point you feel unsure about any aspect of the installation, don't hesitate to consult with or hire professionals. The safety of your family and the protection of your home should always be the top priorities.

By successfully installing a whole-house generator, you're not just preparing for power outages – you're also gaining valuable knowledge about your home's electrical and gas systems. This understanding can be beneficial for future projects and overall home maintenance.

In our next section, we'll discuss gas line safety for homeowners, providing essential knowledge to keep your family safe around gas appliances and systems. Are you ready to become a gas safety expert? Let's move on to this crucial topic!

Chapter 10
Gas Line Safety for Homeowners
Recognizing Gas Odors

Imagine this: You walk into your home and detect an unusual smell. Is it just a passing odor, or could it be a gas leak? Recognizing gas odors is a crucial skill for every homeowner, as it can be your first line of defense against potentially dangerous gas leaks.

When I first became a homeowner, I was unsure about what a gas leak might smell like and how to differentiate it from other household odors. Over time, I've learned that being able to quickly and accurately identify gas odors is an essential safety skill. Let me guide you through the process of recognizing gas odors, sharing insights I've gained to help you protect your home and family.

1. Understanding the Smell of Natural Gas

Natural gas is odorless in its natural state, but gas companies add a distinct odor for safety:

a) The added chemical is usually mercaptan, which has a strong, unpleasant smell
b) Most people describe it as smelling like rotten eggs or sulfur
c) The odor is designed to be noticeable even in small concentrations

Tip: Familiarize yourself with this smell by visiting your gas company's local office, where they often have scratch-and-sniff cards available.

Troubleshooting Tip: If you think you smell gas but aren't sure, don't ignore it. It's better to be cautious and call your gas company than to dismiss a potential leak.

2. Recognizing Propane Gas Odors

If you use propane in your home, be aware of its specific smell:

a) Propane also uses an added odorant, often ethyl mercaptan
b) The smell is often described as similar to skunk spray or rotten cabbage
c) Like natural gas, the odor is designed to be noticeable in small amounts

Tip: If you're new to using propane, ask your propane supplier for a demonstration of the smell.

Troubleshooting Tip: If you notice this smell near your propane tank or appliances, evacuate immediately and call your propane supplier from a safe distance.

3. Identifying Gas Odor Locations

Pay attention to where you smell the odor:

a) Near gas appliances (stoves, water heaters, furnaces)
b) In basements or crawl spaces where gas lines might run
c) Around your gas meter or propane tank
d) In rarely used rooms or closets where leaks might go unnoticed

Tip: Create a mental map of where gas lines and appliances are in your home to help you quickly check these areas.

Troubleshooting Tip: If you smell gas in an unexpected location, don't assume it's not a leak. Gas can travel through walls and floors, so the source may not be immediately apparent.

4. Understanding Odor Intensity

The strength of the gas smell can indicate the severity of the leak:

a) Faint odor: Could be a small leak or residual gas from a recently used appliance
b) Strong odor: Likely indicates a significant leak requiring immediate action
c) Overwhelming odor: Evacuate immediately and call emergency services

Tip: If you're unsure about the intensity, err on the side of caution and treat it as a significant leak.

Troubleshooting Tip: If you notice the smell gets stronger as you move to a specific area, this can help pinpoint the leak's location. However, don't spend time investigating if the smell is strong – evacuate and call professionals.

5. Recognizing Odor Fade

Be aware that in some situations, the gas odor can fade or disappear:

a) In new or seldom-used pipelines, the odorant can stick to the pipe walls
b) Soil can absorb the odorant if there's an underground leak
c) Water in the line can absorb the odorant, masking the smell

Tip: Don't rely solely on smell to detect gas leaks. Use other signs like hissing sounds or bubbles in standing water.

Troubleshooting Tip: If you suspect a leak but don't smell anything, trust your instincts and call your gas company for an inspection.

6. Differentiating from Other Household Odors

Some household smells can be mistaken for gas odors:

a) Sewer gas can smell similar to natural gas
b) Some cleaning products or chemicals can produce sulfur-like smells
c) Certain foods (like eggs) can produce similar odors when spoiled

Tip: If you're unsure, turn off any gas appliances and ventilate the area. If the smell dissipates quickly, it might not be gas.

Troubleshooting Tip: If you frequently smell what you think is gas but professionals find no leak, consider other sources of sulfur-like odors in your home.

7. Training Your Nose

Improve your ability to detect gas odors:

a) Periodically remind yourself of the smell using utility company resources
b) Be aware that your sensitivity to the odor can decrease with prolonged exposure
c) Remember that not everyone smells odors the same way – some people are more sensitive

Tip: Make gas odor recognition part of your family's safety training.

Troubleshooting Tip: If you have allergies or a cold that affects your sense of smell, be extra cautious and consider using a gas detector for added safety.

8. Using Gas Detectors as a Backup

Don't rely solely on your nose:

a) Install gas detectors in your home, especially near gas appliances
b) Choose detectors that can sense both natural gas and propane if you use both
c) Test and maintain your detectors regularly

Tip: Some smart home systems include gas detection features that can alert you even when you're away from home.

Troubleshooting Tip: If your gas detector alarms but you don't smell anything, don't assume it's a false alarm. Evacuate and call for professional help.

9. Educating Family Members

Ensure everyone in your household can recognize gas odors:

a) Teach children what gas smells like and what to do if they smell it
b) Create an evacuation plan in case of a gas leak
c) Post emergency numbers, including the gas company's, in a visible location

Tip: Make gas safety a regular topic of family discussions, especially when seasons change and you start using different gas appliances.

Troubleshooting Tip: If family members report smelling gas at different times, keep a log. This can help identify patterns or intermittent leaks that might be missed during a one-time inspection.

10. Professional Inspections

Even with vigilance, some leaks can be hard to detect:

a) Schedule annual inspections of your gas system by a certified professional
b) Have new gas appliances inspected after installation
c) Request an inspection if you've experienced any recent earthquakes or major construction

Tip: Combine your gas inspection with other annual home maintenance tasks to ensure it's not overlooked.

Troubleshooting Tip: If you've had recurrent small leaks, consider a more comprehensive inspection of your entire gas system. There might be underlying issues causing repeated problems.

Remember, recognizing gas odors is a critical safety skill for any homeowner. While it's important to trust your nose, it's equally important to use other detection methods and err on the side of caution. If you ever suspect a gas leak, don't hesitate to evacuate and call for professional help. The safety of your family and home is always worth the precaution.

By mastering the skill of recognizing gas odors, you're not just protecting your home – you're also gaining peace of mind and the confidence to respond quickly in potentially dangerous situations. In our next section, we'll discuss carbon monoxide awareness, another crucial aspect of gas safety in the home. Are you ready to learn about this silent danger and how to protect against it? Let's move on to this vital topic!

Carbon Monoxide Awareness

Imagine this scenario: You wake up in the middle of the night with a headache and feeling nauseous. Your family members are experiencing similar symptoms. Could it be food poisoning, or is it something more sinister like carbon monoxide (CO) poisoning? Understanding and being aware of the dangers of carbon monoxide is crucial for every homeowner, especially those with gas appliances.

When I first learned about carbon monoxide dangers, I was shocked by how easily it could go undetected. Over time, I've realized that CO awareness is not just about having detectors – it's about understanding the risks, prevention, and quick action. Let me guide you through essential carbon monoxide awareness, sharing insights I've gained to help you protect your home and loved ones from this silent threat.

1. Understanding Carbon Monoxide

Carbon monoxide is a deadly gas that's hard to detect without proper equipment:

a) It's colorless, odorless, and tasteless
b) Produced by the incomplete burning of fuels like gas, oil, coal, and wood
c) Can build up quickly in enclosed spaces

Tip: Remember, you can't rely on your senses to detect CO – that's why detectors are crucial.

Troubleshooting Tip: If you suspect CO but don't have a detector, look for signs like everyone in the house feeling sick at the same time, or pets acting strangely.

2. Recognizing Sources of Carbon Monoxide

Identify potential CO sources in your home:

a) Gas-powered appliances (furnaces, water heaters, stoves, dryers)
b) Fireplaces and wood stoves
c) Vehicles running in attached garages
d) Portable generators used too close to the home

Tip: Create a checklist of all potential CO sources in your home and inspect them regularly.

Troubleshooting Tip: If an appliance starts producing unusual odors or soot, it might be malfunctioning and producing CO. Have it checked immediately.

3. Installing Carbon Monoxide Detectors

Proper placement and maintenance of CO detectors is crucial:

a) Install detectors on every level of your home, including the basement
b) Place detectors near sleeping areas
c) Keep detectors at least 15 feet away from fuel-burning appliances
d) Test detectors monthly and replace batteries annually

Tip: Choose detectors with digital displays to see CO levels even when the alarm isn't sounding.

Troubleshooting Tip: If your detector frequently gives false alarms, check its location. It might be too close to a CO source or affected by extreme temperatures or humidity.

4. Recognizing Symptoms of CO Poisoning

Be aware of the signs of carbon monoxide exposure:

a) Mild exposure: Headache, nausea, dizziness, fatigue
b) Moderate exposure: Severe headache, drowsiness, confusion, rapid heart rate
c) Severe exposure: Unconsciousness, convulsions, cardiac arrest, death

Tip: If multiple people or pets in the house show similar symptoms simultaneously, suspect CO poisoning.

Troubleshooting Tip: If symptoms improve when you leave the house but return when you're back home, this could indicate a CO problem. Don't ignore this pattern.

5. Immediate Actions for Suspected CO Exposure

Know what to do if you suspect carbon monoxide in your home:

a) Evacuate everyone (including pets) immediately
b) Call emergency services from outside the house
c) Don't re-enter until professionals have checked and cleared the area
d) Seek medical attention, even if symptoms are mild

Tip: Develop and practice an evacuation plan with your family, including a designated meeting spot outside.

Troubleshooting Tip: If you can't evacuate through normal exits, break a window to get fresh air and escape. Safety is more important than property damage.

6. Preventing Carbon Monoxide Buildup

Take proactive steps to prevent CO accumulation:

a) Have all fuel-burning appliances inspected annually by a professional
b) Ensure proper ventilation for all appliances and fireplaces
c) Never use outdoor cooking appliances (like grills) indoors
d) Don't leave cars running in attached garages, even with the door open

Tip: Schedule your annual CO prevention inspection at the same time each year, like when you change your clocks for daylight saving time.

Troubleshooting Tip: If you notice excessive moisture or condensation on windows or walls, it could indicate improper ventilation, which increases CO risk. Have your ventilation system checked.

7. Understanding CO Levels

Know what different CO levels mean:

a) 0-9 ppm: Normal background levels
b) 10-29 ppm: Moderate, monitor the situation
c) 30-35 ppm: Relocate to fresh air and call for professional help
d) Above 35 ppm: Evacuate immediately and call emergency services

Tip: Some advanced CO detectors will display these levels. Familiarize yourself with what they mean.

Troubleshooting Tip: If your detector shows low levels consistently, don't ignore it. Even low-level exposure over time can be harmful. Investigate the source.

8. Maintaining Gas Appliances

Proper maintenance is key to preventing CO production:

a) Keep appliances clean and free from debris
b) Ensure burners have a clear, blue flame (yellow or orange indicates a problem)
c) Check vents and flues for blockages regularly
d) Replace old appliances with modern, efficient models

Tip: Create a maintenance schedule for each gas appliance and stick to it rigorously.

Troubleshooting Tip: If you notice a decrease in hot water or heating efficiency, don't just assume it's age-related. It could indicate a combustion problem that's producing CO.

9. Educating Family Members

Ensure everyone in your household understands CO dangers:

a) Teach children about CO and what to do if the detector alarms
b) Show family members how to recognize symptoms of CO poisoning
c) Conduct CO safety drills, similar to fire drills

Tip: Make CO awareness part of your regular family safety discussions.

Troubleshooting Tip: If family members are reluctant to take CO seriously, share real-life stories about CO incidents to emphasize its dangers.

10. Special Considerations

Be aware of situations that might increase CO risk:

a) During power outages if using alternative heating/cooking methods
b) In extreme weather when appliances are working harder
c) When doing home renovations that might affect ventilation
d) In homes with attached garages or fuel-burning appliances in living spaces

Tip: Be extra vigilant about CO during these high-risk times. Use battery-powered CO detectors during power outages.

Troubleshooting Tip: If you're using a generator during a power outage, ensure it's at least 20 feet from your home and not near windows or doors. Many CO poisonings occur during power outages due to improperly placed generators.

Remember, carbon monoxide awareness is about more than just having detectors – it's about understanding the risks, taking preventive measures, and knowing how to respond quickly if CO is detected. By being proactive about CO safety, you're protecting your family from a serious and potentially fatal threat.

Carbon monoxide awareness is a critical part of home safety, especially for homes with gas appliances. By mastering this knowledge, you're not just safeguarding your home – you're potentially saving lives. In our next section, we'll discuss creating a comprehensive home safety plan that incorporates gas and CO safety along with other crucial safety measures. Are you ready to develop a complete safety strategy for your home? Let's move on to this essential topic!

Creating a Home Safety Plan

Imagine this scenario: It's the middle of the night, and you're awakened by the shrill sound of your carbon monoxide detector. In that moment of confusion and fear, do you know exactly what to do? Having a comprehensive home safety plan can mean the difference between panic and confident action in emergencies, especially when it comes to gas-related incidents.

When I first became a homeowner, I underestimated the importance of a detailed safety plan. Over time, I've learned that a well-thought-out plan not only prepares you for emergencies but also gives you peace of mind in your daily life. Let me guide you through the process of creating a comprehensive home safety plan, with a focus on gas and CO safety, sharing insights I've gained to help you protect your home and loved ones.

1. Assessing Your Home's Risks

Start by identifying potential hazards:

a) List all gas appliances and their locations
b) Note areas where CO might accumulate (attached garages, enclosed spaces)
c) Identify potential evacuation obstacles (stairs, windows, locked doors)
d) Consider external factors (weather patterns, nearby industrial sites)

Tip: Walk through your home with fresh eyes, looking for hazards you might overlook daily.

Troubleshooting Tip: If you're unsure about certain risks, consider having a professional home safety inspection to identify less obvious hazards.

2. Creating an Evacuation Plan

Design a clear plan for quickly exiting your home:

a) Draw a floor plan with two exits from each room
b) Mark the location of all gas shut-off valves
c) Designate a family meeting spot outside the home
d) Plan for evacuating pets and assisting family members with special needs

Tip: Use different colored markers on your floor plan to show primary and secondary exit routes.

Troubleshooting Tip: If a room seems to have only one exit, consider purchasing escape ladders for upper floors or creating a "safe room" with a phone and supplies if immediate evacuation is impossible.

3. Installing and Maintaining Safety Devices

Ensure your home is equipped with essential safety devices:

a) Install CO detectors on every level and near sleeping areas
b) Place fire extinguishers in key locations (kitchen, garage, near gas appliances)
c) Consider installing a whole-house gas detector
d) Maintain smoke detectors in all bedrooms and common areas

Tip: Create a maintenance schedule for all safety devices, including battery changes and expiration dates.

Troubleshooting Tip: If a detector frequently gives false alarms, don't disable it. Instead, check its placement or consider upgrading to a more reliable model.

4. Educating Family Members

Ensure everyone knows what to do in an emergency:

a) Teach all family members how to recognize gas odors and CO symptoms
b) Practice your evacuation plan regularly, including different scenarios (day/night, blocked exits)
c) Show everyone how to use fire extinguishers and shut off gas valves
d) Establish a communication plan for when family members are separated

Tip: Make safety education fun for kids by turning it into a game or rewarding them for remembering safety rules.

Troubleshooting Tip: If family members struggle to remember details, create simple, visual reminders to post around the house.

5. Preparing an Emergency Kit

Assemble a kit with essentials for various emergencies:

a) Include water, non-perishable food, first-aid supplies, and medications
b) Add flashlights, batteries, a battery-powered radio, and a phone charger
c) Include copies of important documents (ID, insurance information)
d) Consider special needs items (pet supplies, baby items, etc.)

Tip: Store your emergency kit in an easily accessible location, and review its contents twice a year.

Troubleshooting Tip: If space is an issue, prioritize the most critical items and consider storing some supplies in your car.

6. Establishing a Maintenance Routine

Regular maintenance prevents many safety issues:

a) Schedule annual inspections for all gas appliances
b) Clean dryer vents and chimneys regularly
c) Check for signs of wear on gas lines and connections
d) Keep areas around gas appliances clear of clutter and flammable materials

Tip: Create a yearly home maintenance calendar to ensure no safety checks are overlooked.

Troubleshooting Tip: If you're not comfortable performing certain maintenance tasks, budget for professional services. The cost is worth the safety assurance.

7. Creating an Emergency Contact List

Compile a list of important numbers and information:

a) Include emergency services, utility companies, and your gas provider
b) Add contacts for nearby family or friends who can help in an emergency
c) List important medical information for family members
d) Include the contact information for your insurance company

Tip: Store this list in your phone, but also keep a printed copy in your emergency kit.

Troubleshooting Tip: If you have trouble keeping this information updated, set a recurring reminder on your phone or calendar.

8. Planning for Different Scenarios

Consider various emergency situations:

a) Gas leak during the day vs. night
b) CO buildup during a power outage
c) Fire starting from a gas appliance
d) Natural disasters affecting gas lines (earthquakes, floods)

Tip: Run through "what-if" scenarios with your family to identify any gaps in your plan.

Troubleshooting Tip: If certain scenarios seem particularly challenging, consult with local emergency services for specific advice.

9. Integrating Technology

Use modern technology to enhance your safety plan:

a) Install smart CO and smoke detectors that alert your phone
b) Consider a home security system with gas and CO monitoring
c) Use apps to store and share your emergency plans and contacts
d) Explore community alert systems for local emergencies

Tip: Test any tech-based systems regularly to ensure they're functioning correctly.

Troubleshooting Tip: If you're not tech-savvy, ask a family member or friend to help set up and explain these systems. Don't let technology be a barrier to safety.

10. Regular Review and Updates

Keep your safety plan current:

a) Review and update your plan at least annually
b) Adjust the plan after any home renovations or changes in family structure
c) Stay informed about new safety recommendations or local hazards
d) Incorporate lessons learned from any close calls or actual emergencies

Tip: Set a specific date each year (like New Year's Day) to review your safety plan.

Troubleshooting Tip: If reviewing the entire plan seems overwhelming, break it into smaller tasks throughout the year.

Creating a comprehensive home safety plan is an ongoing process that requires thought, effort, and regular updates. However, the peace of mind and preparedness it provides are invaluable. By focusing on gas and CO safety within your broader home safety strategy, you're addressing some of the most critical and potentially dangerous aspects of home safety.

Remember, the best safety plan is one that everyone in the household understands and can act on instinctively. Regular practice and open discussion about safety will help ensure that in the moment of crisis, everyone knows exactly what to do.

By developing and maintaining a thorough home safety plan, you're not just protecting your property – you're safeguarding what matters most: the lives and well-being of your loved ones. In our next section, we'll discuss legal and insurance considerations related to gas line work and safety. Are you ready to ensure your home is not only safe but also compliant and properly covered? Let's move on to this important aspect of home gas safety!

Chapter 11
Legal and Insurance Considerations
Local Codes and Regulations

Imagine this scenario: You've just completed a DIY gas line installation for your new outdoor kitchen. You're feeling proud of your work, but then a neighbor mentions something about permits and inspections. Suddenly, you're worried – did you inadvertently break any laws or regulations? Understanding and complying with local codes and regulations is crucial for any homeowner dealing with gas lines, not just for legal reasons, but for safety and insurance purposes as well.

When I first started working on gas-related projects in my home, I was overwhelmed by the complexity of local codes and regulations. Over time, I've learned that while navigating these rules can be challenging, it's an essential part of responsible homeownership. Let me guide you through the process of understanding and complying with local codes and regulations for gas line work, sharing insights I've gained to help you stay safe, legal, and protected.

1. Understanding the Importance of Codes and Regulations

Local codes and regulations serve several crucial purposes:

a) Ensure safety standards are met in gas line installations and modifications
b) Protect homeowners from substandard work by unlicensed contractors
c) Maintain consistency in gas system installations across a community
d) Provide legal protection for homeowners in case of accidents or disputes

Tip: View codes as allies in your quest for a safe home, not obstacles to overcome.

Troubleshooting Tip: If a contractor suggests ignoring codes to save money, this is a red flag. Always prioritize safety and compliance over cost savings.

2. Identifying Relevant Authorities

Know which entities govern gas line work in your area:

a) Local building department or permit office
b) State-level regulatory agencies
c) Gas utility companies
d) National organizations like the International Code Council (ICC)

Tip: Create a contact list of these authorities for easy reference when planning projects.

Troubleshooting Tip: If you're getting conflicting information from different sources, ask for written documentation of requirements to clarify discrepancies.

3. Familiarizing Yourself with Key Codes

Common areas covered by gas line codes include:

a) Pipe materials and sizes
b) Installation methods and depths for buried lines
c) Ventilation requirements for gas appliances
d) Safety device installations (like shut-off valves)

Tip: Many local governments now offer online access to their code books. Bookmark these resources for easy reference.

Troubleshooting Tip: If a specific code seems unclear, don't guess. Contact your local building department for clarification before proceeding with work.

4. Obtaining Necessary Permits

Understanding when permits are required:

a) New gas line installations
b) Modifications to existing gas systems
c) Installation of new gas appliances
d) Repairs beyond simple maintenance

Tip: Even if you're hiring a contractor, verify that they've obtained all necessary permits. Ultimately, it's the homeowner's responsibility.

Troubleshooting Tip: If you've already completed work without a permit, don't panic. Many jurisdictions have processes for retroactive permitting. It's better to address this than to leave work unpermitted.

5. Navigating the Permit Process

Steps typically involved in obtaining a permit:

a) Submit detailed plans of your proposed work
b) Pay required fees
c) Schedule and pass inspections at various stages of the project
d) Obtain final approval and certificate of occupancy (if applicable)

Tip: Build permit timelines into your project schedule. Rushing through the permit process can lead to mistakes or oversights.

Troubleshooting Tip: If your permit application is denied, ask for a detailed explanation. Often, simple modifications to your plans can address concerns and lead to approval.

6. Understanding Inspection Requirements

Know what inspectors will be looking for:

a) Proper materials and installation techniques
b) Adherence to approved plans
c) Safety features like proper ventilation and shut-off valves
d) Pressure testing of new or modified gas lines

Tip: Be present during inspections if possible. This is an opportunity to learn and ask questions about your gas system.

Troubleshooting Tip: If you fail an inspection, don't be discouraged. Treat it as a learning experience. Ask the inspector for detailed feedback on what needs to be corrected.

7. Staying Updated on Code Changes

Codes and regulations can change over time:

a) Subscribe to updates from your local building department
b) Attend community meetings where code changes are discussed
c) Join local homeowner associations or groups that share this information
d) Regularly check online resources for code updates

Tip: Set a yearly reminder to review any code changes that might affect your home's gas system.

Troubleshooting Tip: If a recent code change affects your existing system, you may be grandfathered in. Check with local authorities about requirements for updating older installations.

8. Dealing with Older, Non-Compliant Systems

Many older homes have gas systems that don't meet current codes:

a) Understand the concept of "grandfathering" and when it applies
b) Know when upgrades are mandatory (often during major renovations)
c) Consider voluntary upgrades to improve safety and efficiency
d) Document any non-compliant systems for insurance purposes

Tip: Even if your older system is grandfathered, consider upgrading for safety and efficiency benefits.

Troubleshooting Tip: If you're unsure about the compliance status of your older gas system, consider hiring a licensed inspector for a thorough evaluation.

9. Understanding Zoning Laws

Zoning can affect gas line projects, especially for outdoor installations:

a) Check setback requirements for outdoor gas appliances
b) Understand restrictions on propane tank placements
c) Be aware of any historic district regulations that might affect your project
d) Consider neighbor notification requirements for certain projects

Tip: Consult zoning maps and regulations before planning any major gas line projects, especially those visible from the street.

Troubleshooting Tip: If zoning laws seem to prohibit your project, explore variance or special use permit options. Sometimes exceptions can be made with proper application and justification.

10. Keeping Detailed Records

Maintain comprehensive documentation of your gas system:

a) Keep copies of all permits and inspection reports
b) Document any modifications or repairs to your system
c) Save manuals and warranty information for gas appliances
d) Create a maintenance log for your gas system

Tip: Create both physical and digital copies of all important documents related to your gas system.

Troubleshooting Tip: If you're missing documentation for older work, contact your local building department. They often keep records of past permits and inspections.

Understanding and complying with local codes and regulations for gas line work is not just about avoiding fines or legal issues – it's about ensuring the safety of your home and family. While the process can seem daunting, remember that these regulations are in place to protect you and your community.

By familiarizing yourself with local codes, obtaining proper permits, and ensuring all work is inspected and approved, you're not only staying legal but also gaining peace of mind. You'll know that your gas system meets safety standards and that you're protected in case of any issues.

Insurance Requirements

Imagine this scenario: You've just completed a major gas line renovation in your home. Everything seems perfect until a small leak causes damage to your property. You file an insurance claim, only to discover that your policy doesn't cover the incident because the work wasn't properly documented or performed by a licensed professional. Understanding insurance requirements for gas line work is crucial for protecting your home and finances.

When I first delved into the world of home insurance and gas systems, I was surprised by the complexity and specific requirements involved. Over time, I've learned that being proactive about insurance considerations can save you from significant headaches and financial strain. Let me guide you through the key insurance requirements related to gas line work, sharing insights I've gained to help you ensure you're properly covered.

1. Understanding Your Current Coverage

Start by reviewing your existing homeowner's insurance policy:

a) Check for specific clauses related to gas lines and appliances
b) Look for coverage limits on gas-related incidents
c) Understand what types of damage are covered (fire, explosion, water damage from leaks)
d) Note any exclusions related to DIY work or non-licensed contractors

Tip: Create a summary sheet of your gas-related coverage for quick reference.

Troubleshooting Tip: If your policy language is unclear, don't hesitate to contact your insurance agent for clarification. It's better to understand your coverage before an incident occurs.

2. Notifying Your Insurance Company of Changes

Keep your insurer informed about gas system modifications:

a) Report any major gas line installations or modifications
b) Inform them of new gas appliance installations
c) Update them on any changes in the use of gas in your home (e.g., converting to a gas stove)
d) Provide documentation of work completed, including permits and inspections

Tip: Set a reminder to review and update your insurance company annually about any changes to your gas system.

Troubleshooting Tip: If you're unsure whether a change is significant enough to report, err on the side of caution and inform your insurer. Over-communication is better than under-communication in insurance matters.

3. Understanding Liability Coverage

Ensure you have adequate liability protection:

a) Check your policy's liability limits for gas-related incidents
b) Consider increasing liability coverage if you have extensive gas systems
c) Understand how your coverage applies if a gas incident affects neighboring properties
d) Be aware of any liability exclusions for specific types of gas work or appliances

Tip: Consider an umbrella policy for additional liability protection, especially if you have high-value assets.

Troubleshooting Tip: If you're hosting events where you'll be using gas appliances (like outdoor grilling), check if your liability coverage extends to these situations.

4. Documenting Professional Installations

Proper documentation is crucial for insurance coverage:

a) Keep detailed records of all professional gas work
b) Retain copies of contractor licenses and insurance
c) Save all receipts, contracts, and work orders
d) Document before-and-after photos of major gas projects

Tip: Create a digital folder for all gas-related documents, and back it up in cloud storage for easy access.

Troubleshooting Tip: If you've lost documentation for past work, contact the contractors who performed the work. Many keep records and can provide copies of important documents.

5. Understanding DIY Limitations

Be aware of how DIY work affects your coverage:

a) Check if your policy allows for any DIY gas work
b) Understand which tasks must be performed by licensed professionals
c) Know the consequences of DIY work on your coverage and claim validity
d) Consider additional riders or coverage for DIY projects, if available

Tip: Even if you're capable of doing the work, sometimes it's worth hiring a professional to ensure full insurance coverage.

Troubleshooting Tip: If you've already completed DIY work, consider having it inspected and certified by a professional to potentially reinstate full coverage.

6. Regular Maintenance and Its Impact on Insurance

Understand how maintenance affects your coverage:

a) Follow manufacturer-recommended maintenance schedules for gas appliances
b) Keep records of all maintenance and inspections
c) Understand how neglected maintenance can void coverage
d) Consider how professional maintenance might positively impact your premiums

Tip: Create a maintenance calendar and set reminders to ensure you stay on schedule.

Troubleshooting Tip: If you've fallen behind on maintenance, start a catch-up plan immediately. Some insurers may be more lenient if you show a good-faith effort to maintain your system.

7. Understanding Code Compliance and Insurance

Recognize the link between building codes and insurance:

a) Ensure your gas system is up to current codes
b) Understand how non-compliant systems might affect claims
c) Be aware of any grace periods for updating to new code requirements
d) Document any grandfathered systems and how they're viewed by your insurer

Tip: During your annual insurance review, specifically ask about any new code requirements that might affect your coverage.

Troubleshooting Tip: If bringing your system up to code is prohibitively expensive, discuss options with your insurer. Some may offer temporary coverage extensions while you plan for updates.

8. Specialized Coverage for Unique Situations

Consider whether you need additional or specialized coverage:

a) If you work from home, check if your business use affects gas system coverage
b) For historic homes, understand how preservation requirements interact with insurance needs
c) If you have underground gas tanks, look into specific coverage options
d) For homes in disaster-prone areas, consider how this affects gas system coverage

Tip: Consult with an insurance broker who specializes in unique or high-value properties if your situation is complex.

Troubleshooting Tip: If standard insurers won't cover your unique situation, explore surplus line insurers who might offer more flexible policies.

9. Understanding Claims Processes

Be prepared in case you need to file a claim:

a) Know the immediate steps to take in case of a gas-related incident
b) Understand your responsibility to mitigate further damage
c) Be familiar with the documentation required for a claim
d) Know the timelines for filing claims and providing information

Tip: Create an emergency file with all relevant insurance information and contacts for quick access during an incident.

Troubleshooting Tip: If you're unsure about whether to file a claim for a minor incident, consult your insurance agent. Sometimes, it's better to handle small issues out-of-pocket to avoid premium increases.

10. Reviewing and Updating Your Coverage

Regularly assess whether your coverage is still adequate:

a) Review your policy annually, or after any major home upgrades
b) Reassess coverage limits to ensure they match current replacement costs
c) Consider how changes in your financial situation might warrant coverage adjustments
d) Stay informed about new insurance products that might better suit your needs

Tip: Set a specific date each year (like your home purchase anniversary) to review your insurance coverage.

Troubleshooting Tip: If you find your current insurer isn't meeting your needs, don't be afraid to shop around. Different companies may offer better rates or more suitable coverage for your specific situation.

Understanding and meeting insurance requirements for your home's gas system is not just about compliance – it's about protecting your investment and ensuring financial security. While navigating insurance policies can be complex, the peace of mind that comes with proper coverage is invaluable.

By staying informed, maintaining proper documentation, and regularly reviewing your coverage, you're not just fulfilling insurance requirements – you're safeguarding your home and family against potential gas-related incidents. Remember, your insurance policy is a crucial part of your overall home safety strategy.

In our next section, we'll discuss the importance of hiring licensed professionals for gas line work and how to ensure you're working with qualified experts. Are you ready to learn how to choose the right professionals to keep your gas system safe and compliant? Let's move on to this essential aspect of gas line management!

Hiring Licensed Professionals

Imagine this scenario: You're planning a major gas line installation for your new kitchen remodel. You're tempted to hire your neighbor's friend who offers to do the job for a fraction of the price of a licensed professional. It seems like a great deal, but is it worth the risk? Understanding the importance of hiring licensed professionals for gas line work is crucial for your safety, legal compliance, and peace of mind.

When I first faced the decision of hiring for gas line work, I was tempted to cut corners to save money. Over time, I've learned that the expertise and assurance that come with hiring licensed professionals are invaluable. Let me guide you through the process of hiring licensed professionals for gas line work, sharing insights I've gained to help you make informed and safe decisions.

1. Understanding the Importance of Licensing

Recognize why hiring licensed professionals is crucial:

a) Ensures the work meets safety standards and local codes
b) Provides legal protection in case of accidents or damage
c) Often required for insurance coverage and claim validity
d) Guarantees a minimum level of expertise and training

Tip: Think of the license as an insurance policy for the quality and safety of the work.

Troubleshooting Tip: If you've already had work done by an unlicensed individual, consider having it inspected by a licensed professional to ensure safety and compliance.

2. Verifying Licenses

Know how to check a professional's licensing status:

a) Ask for the license number and verify it with your state's licensing board
b) Check the expiration date and ensure it's current
c) Verify that the license covers the specific type of gas work you need
d) Look for any complaints or disciplinary actions against the license holder

Tip: Most state licensing boards have online verification systems for easy checking.

Troubleshooting Tip: If a contractor is hesitant to provide licensing information, consider it a red flag and look for another professional.

3. Understanding Different Types of Licenses

Be aware that there are various licenses related to gas work:

a) Plumbing licenses (often cover gas line installation)
b) HVAC licenses (for heating system work)
c) General contractor licenses (for overall project management)
d) Specialized gas fitter licenses (in some jurisdictions)

Tip: Match the type of license to the specific work you need done.

Troubleshooting Tip: If your project involves multiple systems, you may need professionals with different licenses. Ensure each aspect of the work is covered by the appropriate license.

4. Checking Insurance and Bonding

Verify that the professional is properly insured and bonded:

a) Ask for proof of liability insurance
b) Check for workers' compensation coverage if they have employees
c) Verify bonding, which protects you if the job is not completed
d) Understand the coverage limits and what they mean for your project

Tip: Don't just take their word for it – ask to see current certificates of insurance and bonding.

Troubleshooting Tip: If a contractor's insurance seems inadequate for your project, discuss increasing their coverage or consider finding a better-insured professional.

5. Evaluating Experience and Expertise

Look beyond just the license for true expertise:

a) Ask about years of experience in gas line work
b) Inquire about similar projects they've completed
c) Check for additional certifications or specialized training
d) Ask about their familiarity with local codes and regulations

Tip: Request references from past clients with similar gas line projects.

Troubleshooting Tip: If a contractor has a license but seems inexperienced with your specific type of project, consider looking for someone with more relevant experience.

6. Getting Detailed Quotes and Contracts

Ensure all aspects of the work are clearly defined:

a) Get detailed, written quotes from multiple licensed professionals
b) Ensure the contract specifies all work to be done, materials to be used, and timelines
c) Clarify who is responsible for obtaining permits and scheduling inspections
d) Understand the payment terms and schedule

Tip: Never agree to pay the full amount upfront – a reasonable deposit followed by progress payments is standard.

Troubleshooting Tip: If a quote seems significantly lower than others, be cautious. Ensure they're not cutting corners or using subpar materials.

7. Understanding the Permitting Process

Know how licensed professionals handle permits:

a) Confirm they will obtain all necessary permits
b) Understand how permit costs are handled (included or additional?)
c) Ensure they're familiar with local permitting processes
d) Clarify who will be present for inspections

Tip: Ask to see copies of pulled permits to verify they've been properly obtained.

Troubleshooting Tip: If a contractor suggests skipping permits to save time or money, this is a major red flag. Insist on proper permitting or find another professional.

8. Communicating Effectively

Establish clear communication channels:

a) Discuss preferred methods of communication (phone, email, text)
b) Set expectations for regular updates on project progress
c) Establish a point of contact for questions or concerns
d) Understand their policy on change orders and additional work

Tip: Keep a log of all communications and decisions made throughout the project.

Troubleshooting Tip: If communication breaks down during the project, schedule a face-to-face meeting to address concerns and get back on track.

9. Understanding Warranties and Guarantees

Know what protections you have after the work is complete:

a) Understand the contractor's workmanship warranty
b) Clarify warranties on materials and equipment used
c) Know the process for addressing issues that arise after completion
d) Understand any maintenance requirements to keep warranties valid

Tip: Get all warranty information in writing as part of your contract.

Troubleshooting Tip: If a problem arises after the work is complete, contact the professional promptly. Delays in reporting issues could void warranties.

10. Preparing for the Work

Help ensure a smooth project:

a) Clear the work area and provide necessary access
b) Understand any preparations you need to make (e.g., shutting off gas or water)
c) Clarify work hours and any restrictions (noise, parking, etc.)
d) Discuss how to handle unexpected issues that may arise during the work

Tip: Ask the professional if there's anything specific you can do to facilitate the work.

Troubleshooting Tip: If unexpected issues arise during the project, ask for a clear explanation and get any changes to the scope of work in writing.

Hiring licensed professionals for gas line work is not just about following regulations – it's about ensuring the safety of your home and family. While it might seem more expensive upfront, the peace of mind and long-term safety benefits far outweigh the costs.

Remember, gas line work is not an area where cutting corners is wise. By hiring licensed, insured, and experienced professionals, you're investing in the safety and integrity of your home's gas system.

In our final section, we'll look towards the future of home gas systems, exploring emerging technologies and trends that might affect your gas system in the coming years. Are you ready to glimpse into the future of home gas technology? Let's move on to this fascinating topic!

Chapter 12
Future of Home Gas Systems

Smart Gas Meters

Imagine a world where you never have to worry about estimating your gas bill or wondering about your consumption patterns. Smart gas meters are turning this vision into reality, revolutionizing how we monitor and manage our home gas usage. As we look to the future of home gas systems, smart meters stand out as a game-changing technology that promises to enhance efficiency, safety, and consumer control.

When I first heard about smart gas meters, I was skeptical about their benefits and concerned about potential privacy issues. Over time, I've come to appreciate the advantages they offer and understand how to address common concerns. Let me guide you through the world of smart gas meters, sharing insights I've gained to help you understand and prepare for this exciting technology.

1. Understanding Smart Gas Meters

Smart gas meters are advanced devices that offer several benefits:

a) Provide real-time gas consumption data
b) Eliminate the need for manual meter readings
c) Enable more accurate billing
d) Facilitate better energy management for consumers

Tip: Think of a smart meter as your personal gas consumption assistant, always on duty.

Troubleshooting Tip: If you're concerned about the accuracy of your smart meter, most utility companies offer comparison periods where both smart and traditional meters are monitored to ensure consistency.

2. Real-Time Consumption Monitoring

Smart meters allow you to track your gas usage in real-time:

a) Access consumption data through smartphone apps or web portals
b) Set up alerts for unusual consumption patterns
c) Understand how different appliances affect your gas usage
d) Identify potential leaks or inefficiencies quickly

Tip: Use the real-time data to create a 'energy usage profile' for your home, helping you spot anomalies.

Troubleshooting Tip: If you notice sudden spikes in consumption, check for open windows or doors that might be causing your heating system to work overtime before assuming it's a meter error.

3. Enhanced Billing Accuracy

Smart meters improve the billing process:

a) Eliminate estimated bills
b) Provide more detailed breakdowns of gas usage
c) Enable easier budget planning with predictive usage features
d) Facilitate faster resolution of billing disputes

Tip: Review your first few smart meter bills carefully to understand the new format and information provided.

Troubleshooting Tip: If your bill seems unusually high after smart meter installation, check if there was a long gap between your last traditional reading and the smart meter activation. This might result in a one-time adjustment.

4. Energy Efficiency Improvements

Smart meters can help you reduce your gas consumption:

a) Identify energy-hungry appliances or habits
b) Set personal consumption goals and track progress
c) Receive tailored energy-saving tips based on your usage patterns
d) Participate in energy-saving challenges or programs offered by utility companies

Tip: Use the data from your smart meter to conduct a DIY energy audit of your home.

Troubleshooting Tip: If you're not seeing the energy savings you expected, consider consulting with an energy efficiency expert who can interpret your smart meter data and offer personalized advice.

5. Integration with Smart Home Systems

Smart gas meters can be part of a larger smart home ecosystem:

a) Connect with smart thermostats for optimized heating
b) Integrate with home automation systems for comprehensive energy management
c) Work with smart appliances to schedule usage during off-peak hours
d) Provide data for whole-home energy management apps

Tip: When upgrading appliances, look for "smart-meter compatible" options to maximize integration benefits.

Troubleshooting Tip: If you're having trouble integrating your smart meter with other smart home devices, check if your utility company offers a specific app or platform for compatibility.

6. Improved Leak Detection

Smart meters enhance safety through better leak detection:

a) Identify unusual consumption patterns that might indicate a leak
b) Enable remote shut-off in case of detected leaks (in some advanced systems)
c) Provide early warning for potential safety issues
d) Facilitate faster emergency response from utility companies

Tip: Familiarize yourself with your smart meter's leak detection features and set up alerts if available.

Troubleshooting Tip: If you receive a leak alert, don't ignore it even if you can't smell gas. Verify with your utility company and consider a professional inspection to be safe.

7. Data Privacy and Security

Addressing common concerns about smart meter data:

a) Understand what data is collected and how it's used
b) Learn about your utility's data protection measures
c) Know your rights regarding data sharing and opt-out options
d) Stay informed about regulations protecting smart meter data

Tip: Request your utility company's privacy policy specifically for smart meter data.

Troubleshooting Tip: If you're uncomfortable with data collection, inquire about alternative plans or opt-out options, but be aware this might affect billing accuracy and available features.

8. Installation and Transition Process

Preparing for smart meter installation:

a) Understand the installation timeline and process
b) Know what to expect on installation day
c) Learn how to read and interpret your new smart meter
d) Understand any changes to your billing cycle or format

Tip: Take a photo of your old meter's final reading for your records.

Troubleshooting Tip: If you experience any service interruptions after installation, contact your utility company immediately. Sometimes, reconnection issues can occur.

9. Cost Considerations

Understanding the financial aspects of smart meters:

a) Learn about any upfront costs or fees associated with installation
b) Understand potential long-term savings through better energy management
c) Inquire about new rate plans that might be available with smart meters
d) Consider the cost-benefit of smart appliances that can integrate with the meter

Tip: Ask your utility company for case studies or average savings figures from other customers with smart meters.

Troubleshooting Tip: If you're not seeing the cost savings you expected, review your consumption patterns and consider an energy audit to identify areas for improvement.

10. Future Developments and Upgrades

Staying informed about evolving smart meter technology:

a) Keep an eye on new features and capabilities being developed
b) Understand your utility's upgrade plans for smart meter technology
c) Learn about potential future integrations with renewable energy systems
d) Stay informed about smart grid developments in your area

Tip: Attend local utility company information sessions or webinars about future smart meter plans.

Troubleshooting Tip: If your area is lagging in smart meter adoption, consider forming or joining a local group to advocate for this technology with your utility company and local government.

Smart gas meters represent a significant step forward in home energy management. They offer the potential for greater control over your gas consumption, improved safety through better leak detection, and more accurate billing. While the transition might seem daunting, the long-term benefits in terms of energy efficiency and cost savings can be substantial.

As with any new technology, it's important to stay informed and engaged. Don't hesitate to ask questions, seek clarification from your utility company, and explore the features of your smart meter. By embracing this technology and using it effectively, you can play an active role in managing your home's energy use and contributing to broader energy efficiency goals.

In our next section, we'll explore another exciting development in home gas systems: renewable natural gas. Are you ready to learn about how your gas supply might become more environmentally friendly? Let's dive into this innovative and sustainable approach to natural gas!

Renewable Natural Gas

Imagine a future where the gas heating your home and cooking your meals comes not from fossil fuels, but from sustainable, renewable sources. This isn't science fiction – it's the promise of renewable natural gas (RNG). As we look towards a more sustainable future for home energy, RNG stands out as a promising solution that could revolutionize how we think about natural gas.

When I first heard about renewable natural gas, I was intrigued but skeptical. Could it really be as good as traditional natural gas? Over time, I've come to understand its potential and the challenges it faces. Let me guide you through the world of renewable natural gas, sharing insights I've gained to help you understand this exciting development in home energy.

1. Understanding Renewable Natural Gas

RNG is a sustainable alternative to conventional natural gas:

a) Produced from organic waste materials like food scraps, agricultural waste, and sewage
b) Chemically identical to conventional natural gas, but carbon-neutral
c) Can be used in existing natural gas infrastructure and appliances
d) Offers a way to reduce greenhouse gas emissions from waste and energy production

Tip: Think of RNG as a way to turn waste into energy, closing the loop in our consumption cycle.

Troubleshooting Tip: If you're confused about how RNG differs from biogas, remember that RNG is biogas that has been processed to meet natural gas pipeline quality standards.

2. Sources of Renewable Natural Gas

RNG can be produced from various sources:

a) Landfills: Capturing methane from decomposing waste
b) Wastewater treatment plants: Utilizing sewage sludge
c) Agricultural operations: Using animal manure and crop residues
d) Food waste: Converting discarded food into energy

Tip: Look for local RNG production facilities to understand how it's being made in your area.

Troubleshooting Tip: If you're concerned about the availability of RNG sources in your region, remember that even urban areas can produce significant amounts from wastewater and food waste.

3. Environmental Benefits

RNG offers several environmental advantages:

a) Reduces methane emissions from organic waste
b) Lowers carbon footprint compared to fossil natural gas
c) Provides a use for waste that would otherwise go to landfills
d) Can help cities and states meet renewable energy goals

Tip: Calculate your potential carbon footprint reduction by switching to RNG using online calculators.

Troubleshooting Tip: If the environmental benefits seem too good to be true, look for third-party verified lifecycle analyses of RNG production to understand its full impact.

4. Integration with Existing Infrastructure

One of RNG's biggest advantages is its compatibility with current systems:

a) Can be used in existing natural gas pipelines
b) No need to replace home appliances or heating systems
c) Blends seamlessly with conventional natural gas
d) Allows for gradual transition as RNG production scales up

Tip: Check with your local utility to see if they're already incorporating RNG into their supply.

Troubleshooting Tip: If you're concerned about the performance of your appliances with RNG, rest assured that it meets the same quality standards as conventional natural gas.

5. Cost Considerations

Understanding the economics of RNG:

a) Currently more expensive to produce than conventional natural gas
b) Costs are expected to decrease as production scales up
c) May be offered as a premium option by some utilities
d) Long-term savings potential through avoided emissions and waste management costs

Tip: Look for utility programs that allow you to opt-in to RNG for a portion of your gas usage to manage costs.

Troubleshooting Tip: If the cost seems prohibitive, consider starting with a small percentage of RNG in your mix and gradually increasing as prices become more competitive.

6. Availability and Access

RNG is becoming more widely available:

a) Check if your local utility offers RNG options
b) Look for community biogas projects in your area
c) Understand the current limitations in RNG production and distribution
d) Stay informed about upcoming RNG projects in your region

Tip: Advocate for RNG options with your local utility if they're not currently available.

Troubleshooting Tip: If RNG isn't available in your area yet, consider supporting other renewable energy options in the meantime, like carbon offsets for your natural gas usage.

7. Quality and Safety Considerations

RNG must meet strict quality standards:

a) Processed to remove impurities and match pipeline gas specifications
b) Undergoes regular testing to ensure quality and safety
c) Monitored for consistent energy content (BTU value)
d) Subject to the same safety regulations as conventional natural gas

Tip: If you're curious about RNG quality, ask your utility for their gas quality reports, which should include RNG if it's in the mix.

Troubleshooting Tip: If you notice any changes in your gas appliance performance after RNG introduction, contact your utility. While unlikely, it's important to report any issues promptly.

8. Future Developments and Scaling

The RNG industry is rapidly evolving:

a) New technologies are improving production efficiency
b) Increasing number of RNG production facilities coming online
c) Growing interest from utilities and energy companies
d) Potential for hydrogen blending with RNG in the future

Tip: Keep an eye on energy industry news for developments in RNG technology and production.

Troubleshooting Tip: If progress seems slow in your area, remember that building RNG infrastructure takes time. Stay engaged with local energy discussions to understand the timeline for adoption.

9. Policy and Incentives

Government policies play a crucial role in RNG adoption:

a) Look for state or federal incentives for RNG production and use
b) Understand renewable energy credits and how they apply to RNG
c) Stay informed about local and national clean energy goals
d) Participate in public discussions about renewable energy policies

Tip: Attend local government meetings or energy forums to learn about and contribute to RNG policy discussions.

Troubleshooting Tip: If policies seem to be hindering RNG adoption in your area, consider joining or forming advocacy groups to push for supportive legislation.

10. Home Energy Planning with RNG

Incorporating RNG into your home energy strategy:

a) Consider RNG when planning long-term home energy upgrades
b) Explore hybrid systems that combine RNG with other renewable sources
c) Understand how RNG fits into overall home energy efficiency efforts
d) Consider the impact of RNG availability on decisions like gas vs. electric appliances

Tip: Create a long-term home energy plan that includes the potential for increasing RNG use over time.

Troubleshooting Tip: If you're unsure how to balance RNG with other energy options, consider consulting an energy advisor who can help create a comprehensive plan for your home.

Renewable Natural Gas represents an exciting development in the quest for more sustainable home energy solutions. It offers the potential to significantly reduce the carbon footprint of our gas usage while utilizing waste products that would otherwise contribute to environmental problems. While RNG is still in the early stages of widespread adoption, its compatibility with existing infrastructure makes it a promising option for a cleaner energy future.

As with any emerging technology, it's important to stay informed and engaged. Don't hesitate to ask questions of your utility providers, participate in local energy discussions, and consider how RNG might fit into your long-term home energy plans. By understanding and supporting the development of RNG, you can play a part in shaping a more sustainable future for home gas systems.

In our final section, we'll explore hybrid systems that combine various energy sources, including RNG, to create more resilient and efficient home energy solutions. Are you ready to learn about the cutting-edge of home energy technology? Let's dive into this exciting and complex topic!

Hybrid Systems

Imagine a home where your energy needs are met through a seamless integration of multiple sources - natural gas, electricity, solar power, and even stored energy. This is the promise of hybrid systems, a cutting-edge approach to home energy management that combines the best of various technologies to optimize efficiency, reliability, and sustainability.

When I first encountered the concept of hybrid systems, I was both excited and overwhelmed by the complexity. Over time, I've come to appreciate the flexibility and resilience these systems offer. Let me guide you through the world of hybrid energy systems, sharing insights I've gained to help you understand and potentially implement this advanced approach to home energy.

1. Understanding Hybrid Systems

Hybrid systems combine multiple energy sources and technologies:

a) Integrate renewable sources (like solar) with traditional ones (like natural gas)
b) Often include energy storage solutions (batteries or thermal storage)
c) Use smart controllers to optimize energy use based on availability and demand
d) Can switch between energy sources to maximize efficiency and minimize costs

Tip: Think of a hybrid system as an energy orchestra, with each component playing its part at the right time for optimal performance.

Troubleshooting Tip: If you're overwhelmed by the complexity, start by focusing on understanding how two energy sources could work together in your home before considering more complex systems.

2. Components of a Hybrid System

Common elements in a residential hybrid system include:

a) Natural gas or propane-powered appliances (furnace, water heater, stove)
b) Solar panels for electricity generation
c) Battery storage system
d) Smart energy management system
e) Heat pump for efficient heating and cooling
f) Potential for integrating renewable natural gas

Tip: Create a diagram of your ideal hybrid system to visualize how different components would interact.

Troubleshooting Tip: If certain components seem incompatible, consult with an energy systems specialist. They can often suggest solutions or alternatives to make your system work.

3. Benefits of Hybrid Systems

Hybrid systems offer numerous advantages:

a) Increased energy independence and resilience
b) Potential for significant cost savings over time
c) Reduced carbon footprint through optimized energy use
d) Ability to take advantage of time-of-use pricing from utilities
e) Flexibility to adapt to changing energy needs and technologies

Tip: Calculate potential savings and environmental impact using online hybrid system calculators.

Troubleshooting Tip: If initial cost estimates seem prohibitive, remember that hybrid systems can often be implemented in stages. Start with the most impactful components for your situation.

4. Integrating Natural Gas in Hybrid Systems

Natural gas plays a crucial role in many hybrid setups:

a) Provides reliable backup for renewable sources
b) Can be used for high-demand applications like heating and cooking
c) Works well in conjunction with solar for comprehensive energy coverage
d) Potential for using renewable natural gas to further reduce environmental impact

Tip: When planning your hybrid system, consider which applications are best suited for gas versus electricity in your home.

Troubleshooting Tip: If you're concerned about relying on gas, remember that modern gas appliances are highly efficient and can complement renewable sources effectively.

5. Smart Energy Management

The brain of a hybrid system is its energy management technology:

a) Uses AI and machine learning to optimize energy use
b) Can predict energy needs based on usage patterns and weather forecasts

c) Automatically switches between energy sources for maximum efficiency

d) Provides detailed energy usage data and suggestions for improvement

Tip: Look for energy management systems that offer user-friendly interfaces and mobile apps for easy monitoring and control.

Troubleshooting Tip: If your energy management system seems to make unexpected decisions, check if it's fully updated and properly configured for your specific setup and preferences.

6. Energy Storage Solutions

Storage is key to maximizing the benefits of a hybrid system:

a) Battery systems store excess solar energy for use during peak times or outages

b) Thermal storage (like hot water tanks) can store energy in the form of heat

c) Potential for using electric vehicles as additional storage in vehicle-to-home setups

d) Compressed air or hydrogen storage for long-term energy banking

Tip: Size your storage system based on your typical daily energy use and the duration of backup power you desire.

Troubleshooting Tip: If your battery storage isn't performing as expected, check for issues like improper installation, outdated firmware, or degraded battery cells.

7. Cost Considerations

Understanding the financial aspects of hybrid systems:

a) Higher upfront costs compared to traditional single-source systems
b) Potential for significant long-term savings on energy bills
c) Availability of tax incentives and rebates for renewable energy components
d) Importance of considering lifetime costs and savings, not just initial investment

Tip: Create a detailed cost-benefit analysis over a 10-20 year period to fully understand the financial implications.

Troubleshooting Tip: If the payback period seems too long, explore options for phased implementation or focus on the most cost-effective components first.

8. Installation and Maintenance

Implementing a hybrid system requires careful planning:

a) Work with experienced professionals familiar with integrated energy systems
b) Ensure all components are compatible and properly sized for your needs
c) Plan for regular maintenance of all system components
d) Consider future expandability when designing your system

Tip: Create a comprehensive maintenance schedule for each component of your hybrid system.

Troubleshooting Tip: If you experience issues post-installation, start by checking the connections and communications between different system components. Often, integration problems are at the root of hybrid system issues.

9. Adapting to Changing Technologies

Hybrid systems should be flexible enough to incorporate new technologies:

a) Design with modularity in mind to allow for easy upgrades
b) Stay informed about emerging energy technologies
c) Consider the potential for integrating electric vehicle charging
d) Be open to adding new renewable sources as they become viable

Tip: Regularly review your system's performance and new technology options to identify potential upgrades.

Troubleshooting Tip: If a new component isn't integrating well with your existing system, check for software updates or consider upgrading your energy management system to a more compatible version.

10. Regulatory and Utility Considerations

Navigate the regulatory landscape for hybrid systems:

a) Understand local building codes and permit requirements
b) Check with your utility about grid connection policies for hybrid systems
c) Explore net metering options if you plan to sell excess energy back to the grid
d) Stay informed about changes in energy policies that might affect hybrid systems

Tip: Develop a relationship with your local utility's renewable energy department for ongoing support and information.

Troubleshooting Tip: If you encounter regulatory barriers, look for local renewable energy advocacy groups that can provide guidance and support in navigating complex regulations.

Hybrid systems represent the cutting edge of home energy technology, offering a flexible, efficient, and sustainable approach to meeting our energy needs. While they can be complex, the benefits in terms of energy independence, cost savings, and environmental impact make them an attractive option for forward-thinking homeowners.

As with any advanced technology, it's crucial to do thorough research, work with experienced professionals, and stay informed about new developments. By carefully planning and implementing a hybrid system, you can create a home energy solution that's not only efficient and cost-effective but also ready for the future of energy.

Remember, the world of energy is rapidly evolving, and hybrid systems are at the forefront of this change. By embracing this technology, you're not just optimizing your home's energy use – you're participating in the broader transition to a more sustainable and resilient energy future.

Conclusion

As we conclude our journey through the world of gas line systems, from the basics of installation to the cutting-edge technologies shaping our energy future, it's clear that this field is both complex and rapidly evolving. Whether you're a homeowner looking to understand your gas system better, a DIY enthusiast considering a gas-related project, or simply someone interested in the future of home energy, the knowledge you've gained here provides a solid foundation for making informed decisions about your home's gas infrastructure.

We've covered a wide range of topics, from the critical safety considerations of working with gas lines to the exciting possibilities of renewable natural gas and hybrid energy systems. Throughout this book, we've emphasized the importance of safety, proper planning, and adherence to local codes and regulations. Remember, when it comes to gas systems, there's no substitute for caution and expertise.

As we look to the future, it's evident that gas systems will continue to play a significant role in our homes, albeit in increasingly innovative and sustainable ways. The integration of smart technologies, the rise of renewable natural gas, and the development of hybrid systems all point to a future where our energy use is more efficient, environmentally friendly, and tailored to our individual needs.

However, with these advancements come new challenges and responsibilities. Staying informed about evolving technologies, understanding the implications of our energy choices, and actively participating in the shift towards more sustainable energy solutions are becoming increasingly important for homeowners.

Whether you're maintaining an existing gas system, planning an upgrade, or considering a transition to newer technologies, the key is to approach each decision with careful consideration and a commitment to safety and efficiency. Don't hesitate to seek professional advice when needed, stay current with local regulations, and always prioritize the safety of your home and family.

Remember, your home's energy system is more than just pipes and appliances – it's a crucial part of creating a comfortable, efficient, and sustainable living environment. By understanding and optimizing your gas system, you're not just improving your own home; you're contributing to a broader movement towards smarter, more responsible energy use.

As you move forward, take the knowledge you've gained here and use it as a springboard for further learning and informed decision-making. The world of home energy is dynamic and full of opportunities for those willing to engage with it thoughtfully and proactively.

Thank you for joining us on this exploration of gas line systems and home energy solutions. Here's to safe, efficient, and sustainable homes, now and in the future!

www.ingramcontent.com/pod-product-compliance
Lightning Source LLC
Chambersburg PA
CBHW071912210526
45479CB00002B/386